Elements

SULPHUR

S

How to use this book

This book has been carefully developed to help you understand the chemistry of the elements. In it you will find a systematic and comprehensive coverage of the basic qualities of each element. Each two-page entry contains information at various levels of technical content and language, along with definitions of useful technical terms, as shown in the thumbnail diagram to the right. There is a comprehensive glossary of technical terms at the back of the book, along with an extensive index, key facts, an explanation of the Periodic Table, and a description of how to interpret chemical equations.

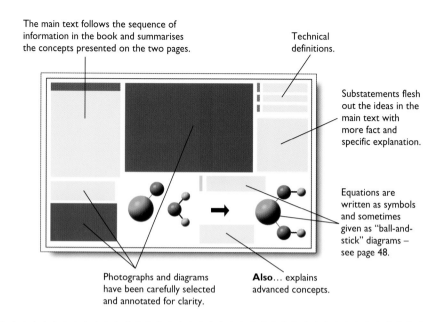

The main text follows the sequence of information in the book and summarises the concepts presented on the two pages.

Technical definitions.

Substatements flesh out the ideas in the main text with more fact and specific explanation.

Equations are written as symbols and sometimes given as "ball-and-stick" diagrams – see page 48.

Photographs and diagrams have been carefully selected and annotated for clarity.

Also… explains advanced concepts.

· ·

An Atlantic Europe Publishing Book

Author
Brian Knapp, BSc, PhD
Project consultant
Keith B. Walshaw, MA, BSc, DPhil
        (Head of Chemistry, Leighton Park School)
Industrial consultant
Jack Brettle, BSc, PhD (Chief Research Scientist, Pilkington plc)
Art Director
Duncan McCrae, BSc
Editor
Elizabeth Walker, BA
Special photography
Ian Gledhill
Illustrations
David Woodroffe and David Hardy
Designed and produced by
EARTHSCAPE EDITIONS
Print consultants
Landmark Production Consultants Ltd
Reproduced by
Leo Reprographics
Printed and bound by
Paramount Printing Company Ltd

Suggested cataloguing location
Knapp, Brian
        Sulphur
        ISBN 1 869860 84 5
        – *Elements* series
540

Acknowledgements
The publishers would like to thank the following for their kind help and advice: *Steve Rockell, John Chrobnick, Tim Fulford, ICI (UK) and Rolls-Royce plc.*

Picture credits
All photographs are from the **Earthscape Editions** photolibrary except the following:
(c=centre t=top b=bottom l=left r=right)
courtesy of **ICI(UK)** 28b; courtesy of **Rolls-Royce plc** 23t; **USGS** 8/9t and **ZEFA** 16/17t, 40/41, 16/17

Front cover: A spectacular sample of amber-coloured rhombic crystals of sulphur from Poland.
Title page: Sulphur burns in oxygen with a characteristic blue flame.

First published in 1996 by
Atlantic Europe Publishing Company Limited, Greys Court Farm,
Greys Court, Henley-on-Thames, Oxon, RG9 4PG, UK.

This product is manufactured from sustainable managed forests. For every tree cut down at least one more is planted.

The demonstrations described or illustrated in this book are not for replication. The Publisher cannot accept any responsibility for any accidents or injuries that may result from conducting the experiments described or illustrated in this book.

Contents

Introduction

An element is a substance that cannot be broken down into a simpler substance by any known means. Each of the 92 naturally occurring elements is therefore one of the fundamental materials from which everything in the Universe is made. This book is about the element sulphur.

Sulphur

Sulphur is a bright yellow, tasteless solid and a very reactive element. It is found in a wide range of minerals and is one of the products of a volcanic eruption. Perhaps this is why many people of previous centuries associated sulphur (also known as brimstone) with the unpleasant afterlife known as Hell. In the New Testament, Hell is described as a "lake that burns with fire and brimstone".

The pure element sulphur has always been thought to have strange properties. A spinning ball of it was used in one of the world's first demonstrations of static electricity. It was found that when the ball was touched by a hand, the ball began to glow.

The element sulphur is a non-metal and will not dissolve in water. The pure element sulphur has very little smell. The smell you might associate with sulphur – bad eggs – is actually a compound of sulphur, the gas known as hydrogen sulphide. Sulphur compounds are also responsible for the smell in garlic, mustard, onions and cabbage. A sulphur compound even gives skunks their ferociously powerful and long-lasting smell.

Indeed, sulphur is a part of all living tissues. Sulphur is fixed into proteins in plants, and acquired by animals who eat the plant materials.

Despite all of the unfortunate connections, sulphur has long had a beneficial medicinal role. It was used both externally, in the form of ointments for the skin and vapours to fumigate diseased places, or internally as the medicine called brimstone. "Brimstone and treacle" was commonly used in Victorian times, and was made famous in the stories of Charles Dickens. In the modern world, the group of drugs known as sulphonamides are used as antimicrobials, one of the more important groups of medicines available today to cure infections of the digestive system.

Because sulphur occurs in all living things, it may be concentrated as tissues decay. This is why sulphur is a common (and unwelcome) component of coal, oil and, to a lesser extent, natural gas. By burning these fossil fuels, sulphur forms a range of gases, including sulphur dioxide, which may cause acid rain.

Sulphur dioxide is an important gas. As well as forming acid when dissolved in water, it is a bleaching agent used in many industrial processes.

The main use of sulphur in large volumes is to produce sulphuric acid, a major starting material in the production of many fertilisers.

◀ Crystals of amber-coloured orthorhombic sulphur set on a rock groundmass. The other common variety, monoclinic sulphur, is shown on page 10.

Properties of sulphur

Native, or pure, sulphur is a soft, yellow, crumbly material. By heating it, many of the special properties of this element are clearly seen.

► The "buckled" ring structure of a molecule of one form of sulphur, as seen from the side and above.

❶► Sulphur is usually found in a laboratory as a crumbly yellow powder. This is the starting material for the demonstrations on this page.

❷◄ Sulphur melts at 115°C. Despite this low melting temperature, it takes a long while for all of the sulphur to melt because sulphur is a poor conductor of heat.

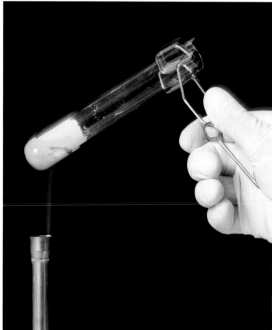

❸► Molten sulphur is an amber liquid that is quite runny (mobile). The reason for this is that sulphur atoms clump into "buckled" rings, each containing eight atoms. The energy of heating allows the rings to spread apart enough to slide over each other.

To imagine how this works, think of a can of spaghetti-rings. When cold, the contents of the emptied can will stand up in a saucepan (they act as a solid), but when heated, the rings start to slide around.

4▶ When more heat is applied the sulphur darkens and the liquid becomes stickier (more viscous). When it reaches 187°C, the tube can be turned over and the liquid will not move. The extra heating has ruptured the rings and they have formed into chains that are now entangled. (Compare the entangling to strands of spaghetti that have been stirred vigorously.)

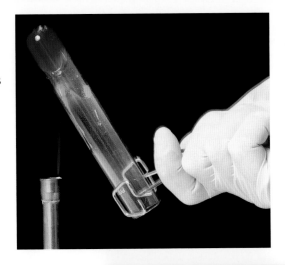

5▶ Heat even further (to 444°C) and the sulphur turns black. The liquid becomes mobile again because it has reached its boiling point. The extra heat energy has ruptured the chains and they now lie in short lengths that can easily move about. (Compare this to chopping up the strands of spaghetti.)

6▼ The liquid can now be poured into cold water. Pick it out of the water and it can be pulled about like plastic.

The "crash-cooling" has taken energy away quickly, causing the sulphur to form enormously long chains. As it is moulded, it gets back some energy and returns to rings, gradually turning back to a solid.

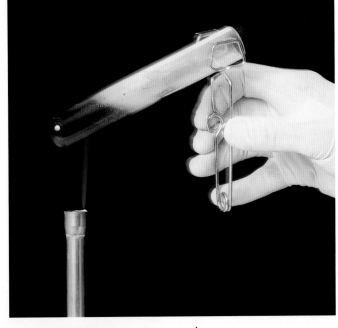

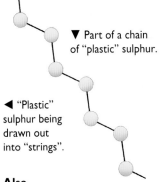

▼ Part of a chain of "plastic" sulphur.

◀ "Plastic" sulphur being drawn out into "strings".

Also...
When the vapours of boiling sulphur are cooled, they condense to form "flowers of sulphur", a mixture of several kinds of sulphur.

The origin of sulphur

Sulphur is thought to have been made in the stars as enormous temperatures and pressures caused elements of lower atomic weight to fuse together.

Native sulphur is formed in many environments associated with volcanic activity, and the characteristic bright yellow sulphur crystals can be seen in many rocks surrounding volcanoes as well as in the rocks formed near geysers. Sulphur is found in most of the Earth's rocks, often as a sulphide, as well as in small amounts in all living tissue. It is also associated with salt domes.

Undersea volcanic activity

Many of the world's volcanoes erupt under the deep oceans along lines where the Earth's crustal plates split apart. At these places water seeps into the rocks and combines with the sulphur gases rising from the magma below the ocean floor. This produces eruptions of superheated acidified water which can also contain dissolved elements such as iron, zinc and copper. Flows of such water are called hydrothermal vents.

As the water flows out into the cold ocean, the cooling hydrothermal water can hold less and less sulphur in solution, and the sulphur starts to crystallise out. Over time, vents can become the sources of vast deposits of sulphur compounds called sulphides. The sulphides are compounds with other elements such as copper and iron, and as such provide potentially vital sources of metal ore.

Also...

In the deep reaches of the ocean, a wide range of living creatures depend on sulphur for life. These places have little oxygen, but a wide variety of sulphur-using bacteria have evolved that make their tissues using sulphur instead of oxygen. These form the basis of a food chain that includes worms and clams.

◄ The crater of Mt Pinatubo. Sulphur can help predict an eruption. When a volcano is about to erupt, it produces more of the gases that will eventually burst forth, such as sulphur dioxide. Detecting an increase in sulphur dioxide emissions was one of the factors that led to the prediction of the Mt Pinatubo eruption in the Philippines in 1991 and the eruption of Mt Unzen in Japan.

▶ **Venus**
Sulphur is common on other planets in the Solar System besides Earth. For example, the main materials making up the atmosphere of the planet Venus are sulphuric acid droplets and particles of solid sulphur.

hydrothermal: a process in which hot water is involved. It is usually used in the context of rock formation because hot water and other fluids sent outwards from liquid magmas are important carriers of metals and the minerals that form gemstones.

ore: a rock containing enough of a useful substance to make mining it worthwhile.

sulphide: a sulphur compound that contains no oxygen (e.g. iron sulphide (FeS)).

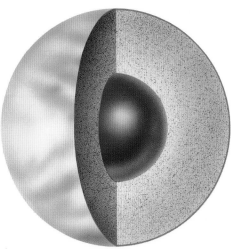

Sulphur and salt

Salt (sodium chloride) deposits are among the most common materials left behind after the evaporation of lakes in deserts. But along with the salt, other minerals are deposited, including calcium sulphate (see page 13). It is thought that sometimes the calcium sulphate was decomposed by bacteria, forming limestone and releasing native sulphur, which then formed veins within cracks in the limestone. This is a very valuable, concentrated resource, which can be melted out of the limestone using the Frasch process (described on page 17). Suitable domes of sulphur have been found in the rocks below the Gulf of Mexico, the Red Sea and Germany.

◄ Pure sulphur collects on rocks near the geysers of Rotorua, New Zealand.

Crystals of sulphur

Sulphur provides striking and colourful crystals. All crystals form in one of seven categories, known as the crystal systems. Some elements, such as sulphur, can form crystals in more than one crystal system.

Sulphur produces bright yellow crystals in the monoclinic system (where crystals look like double-ended chisel blades) and amber crystals in the rhombic system (where crystals look like three-dimensional parallelograms, like a matchbox whose base has been fixed while its top has been pushed sideways).

Brimstone is also pure sulphur, but it is not in crystalline form. It is formed as bacteria consume hydrogen sulphide gas in environments with no oxygen.

▲ To produce these beautiful chisel-like monoclinic crystals, sulphur was dissolved in hot methyl benzene and the dissolved yellow solution allowed to cool. A hot solution can contain more resulting material than a cool solution, so as the solution cools, it reaches a temperature at which the solution cannot contain any more sulphur (it is a saturated solution) and some of the sulphur begins to form into a solid. The growing crystals act as a focus for the deposition of more sulphur in the cooling liquid, so the crystals continue to grow.

crystal systems: seven patterns or systems into which all of the world's crystals can be grouped. They are: cubic, hexagonal, rhombohedral, tetragonal, orthorhombic, monoclinic and triclinic.

cubic crystal system: groupings of crystals that look like cubes.

monoclinic system: a grouping of crystals that look like double-ended chisel blades.

saturated: a state where a liquid can hold no more of a substance. If any more of the substance is added, it will not dissolve.

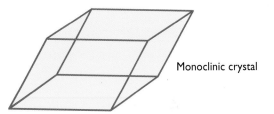

Monoclinic crystal

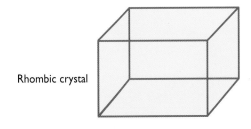

Rhombic crystal

▲ ◄ ◄ ▼ Sulphur crystals form in two of the crystal systems, monoclinic and rhombic. Those to the left are rhombic crystals of sulphur. Another way to identify them is by their colour: rhombic crystals have an amber colouring; monoclinic crystals (such as those on the far left and below) are bright yellow.

Minerals containing sulphur

Sulphur, being a reactive element, occurs in combination with a wide range of other elements. Most of these compounds are sulphides, often known as pyrites. If the compounds also include oxygen, they are known as sulphates.

Most sulphides feel heavy because they contain heavy sulphur atoms densely packed inside the minerals. Most of the sulphides are either brassy-coloured or dark, and some are very distinctive. For example, cadmium sulphide is bright orange (page 25).

Pyrite (iron sulphide), is perhaps the best known of the sulphides. Pyrite can be used as an iron ore (a rock that contains useful amounts of metal). Other main sulphur ores include galena (lead sulphide), sphalerite (zinc sulphide) and chalcopyrite (copper sulphide).

Many of the world's most important ore bodies are sulphides. The huge open-cast mine at Bingham, Utah, USA – claimed to be the world's largest man-made hole – has been dug into veins of copper sulphide. The molybdenum mine at Climax, near Leadville, Colorado, USA, is also made from many veins of sulphide ore. When at its peak, it produced 80% of the entire world's supply of molybdenum. In Sudbury, Ontario, Canada, there are also large deposits of copper and nickel sulphides.

Sulphur is also found in calcium sulphate (as gypsum – used for Plaster of Paris and wall boards) and barium sulphate (as the very heavy mineral barite).

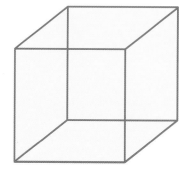

▲ Many crystals can form part of the cubic crystal system. Pyrite normally forms simple cubes, as shown above.

▲ Lead sulphide, or galena, is a common mineral, found in places that have experienced volcanic activity. It is associated with zones from which miners collect other ores such as silver and tin. It is easily spotted by its dark grey cubic (box-shaped) crystals. It is also found in limestone and dolomite rocks through which heated waters have passed.

Galena is a soft mineral. When it is rubbed against a surface it leaves a grey streak of colour, showing its lead content. It is a very heavy mineral and the main source of lead for the world's industries.

Pyrite

This is the most common mineral containing sulphur. It has a brassy colour that resembles gold, and for this reason it has been called "fool's gold".

Some prospectors were fooled into believing they had found a rich vein of gold, when they had found a vein containing pyrite instead. It is, however, much less dense than gold, and much harder. The darker form of pyrite is called marcasite.

Pyrite will react with oxygen in moist air so that the bright surface of a newly exposed pyrite quickly becomes dull and then changes to a fine grey powder.

Pyrite is found abundantly in connection with hot volcanic waters on the ocean floor, but is also formed close to places where the liquids associated with volcanic eruptions force their way into the surrounding rocks.

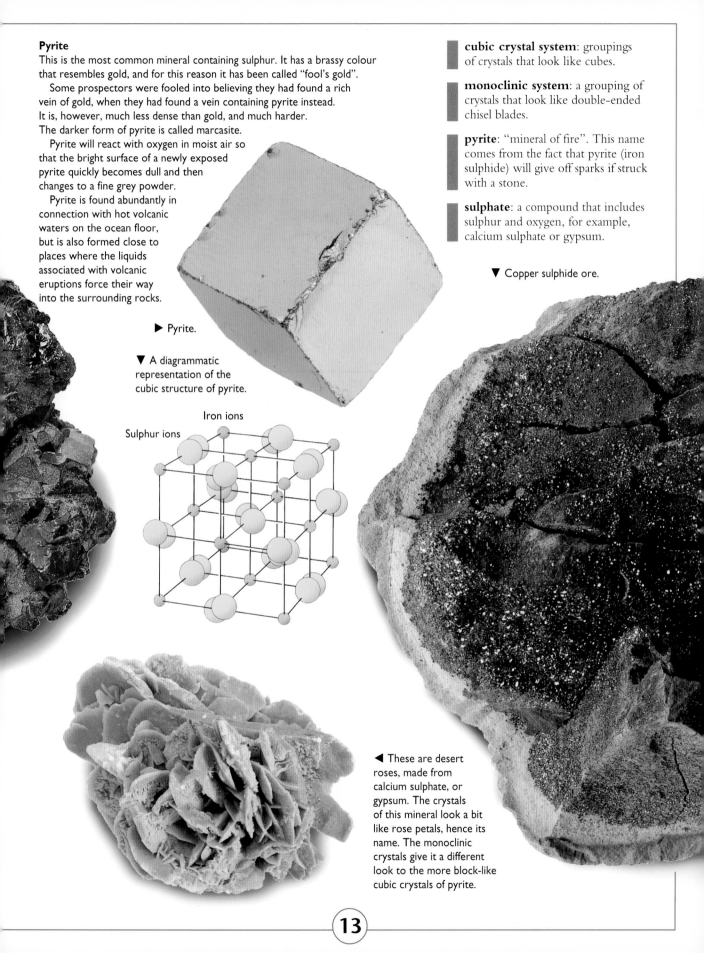

cubic crystal system: groupings of crystals that look like cubes.

monoclinic system: a grouping of crystals that look like double-ended chisel blades.

pyrite: "mineral of fire". This name comes from the fact that pyrite (iron sulphide) will give off sparks if struck with a stone.

sulphate: a compound that includes sulphur and oxygen, for example, calcium sulphate or gypsum.

▼ Copper sulphide ore.

► Pyrite.

▼ A diagrammatic representation of the cubic structure of pyrite.

Iron ions

Sulphur ions

◄ These are desert roses, made from calcium sulphate, or gypsum. The crystals of this mineral look a bit like rose petals, hence its name. The monoclinic crystals give it a different look to the more block-like cubic crystals of pyrite.

Reactivity of sulphur

Sulphur is one of the most reactive of all the elements. This is why sulphur will react with most metals (in the absence of air) to form sulphides and why many sulphur compounds are sulphides.

Sulphur will also react with oxygen in the air (a piece of iron sulphide left lying about will eventually change to iron sulphate) and will readily burn in oxygen to produce sulphur dioxide gas.

In this demonstration, some sulphur powder is placed on a suitable metal holder and ignited in air. It is then introduced into a gas jar filled with oxygen, where it burns with a blue flame.

❶▶ Some powdered sulphur, oxygen in a gas jar and a deflagrating spoon (a metal spoon in which sulphur is burned). Notice that the deflagrating spoon has a metal disc above it in order to protect anyone holding the spoon while the sulphur is burning. The disc also serves as a support for holding the spoon on the middle of the gas jar (as to the right).

Also...

Deflagration is the old word for combustion, hence "deflagrating spoon".

EQUATION: Burning sulphur in air

Sulphur + oxygen ⇨ sulphur dioxide

$$S(s) \quad + \quad O_2(g) \quad ⇨ \quad SO_2(g)$$

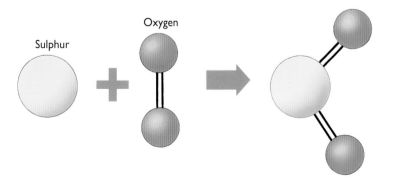

Sulphur   Oxygen

> **combustion**: the special case of oxidisation of a substance where a considerable amount of heat and usually light are given out. Combustion is often referred to as "burning".

❷◄ The sulphur is ignited in air and then introduced to the oxygen, where it burns with a bright blue flame.

Burning sulphur compounds

When sulphur or sulphur-containing compounds are burned, the reaction always produces sulphur dioxide gas as the example equation below shows. This can be a major problem for the iron and copper-refining industries as well as for power stations burning sulphur-rich coal or oil.

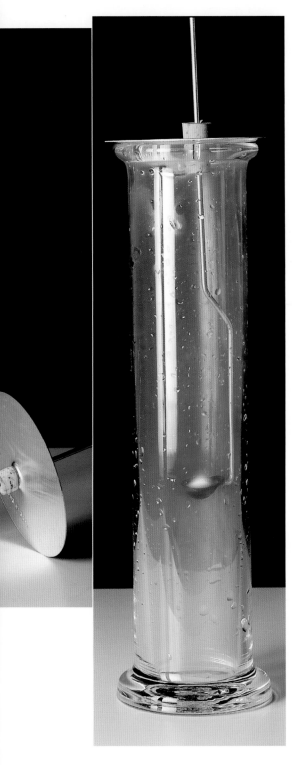

EQUATION: Burning sulphide ore

Iron sulphide + oxygen ⇨ iron oxide + sulphur dioxide

$$4FeS(s) \quad + \quad 7O_2(g) \quad ⇨ \quad 2Fe_2O_3(s) \quad + \quad 4SO_2(g)$$

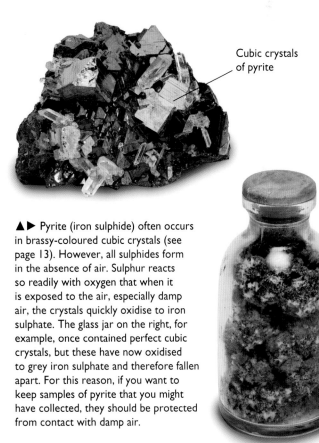

Cubic crystals of pyrite

▲▶ Pyrite (iron sulphide) often occurs in brassy-coloured cubic crystals (see page 13). However, all sulphides form in the absence of air. Sulphur reacts so readily with oxygen that when it is exposed to the air, especially damp air, the crystals quickly oxidise to iron sulphate. The glass jar on the right, for example, once contained perfect cubic crystals, but these have now oxidised to grey iron sulphate and therefore fallen apart. For this reason, if you want to keep samples of pyrite that you might have collected, they should be protected from contact with damp air.

EQUATION: Weathering of pyrite (iron sulphide) in damp air

Iron sulphide + water + oxygen ⇨ iron sulphate + sulphuric acid

$$2FeS_2(s) \quad + \quad 2H_2O(l) \quad + \quad 7O_2(g) \quad ⇨ \quad 2FeSO_4(s) \quad + \quad 2H_2SO_4(aq)$$

Extracting sulphur

For centuries it was commonplace to collect sulphur by lowering people down the inside of volcanoes in baskets so they could scrape the sulphur off the walls of the vents. Needless to say, this was not a very popular occupation.

The main technique used to recover sulphur from buried deposits of native sulphur, such as those associated with salt domes (see pages 8 and 9), relies on the fact that sulphur has a low melting point, while at the same time being insoluble in water.

Superheated water (raised to a temperature of about 165°C) is pumped underground through a pipe. Inside this pipe are two smaller pipes. Compressed air is pumped down the central pipe, and a frothy mixture of liquid sulphur, water and air is pushed up through the remaining pipe. This system is called the Frasch process, named for its inventor Herman Frasch, an American chemist who invented the process in 1891.

Most sulphur mined by the Frasch process is transported as liquid in insulated railway wagons and ships and taken to sulphuric acid plants.

Sulphur from pyrites
In the past, many sulphide-rich ores have been roasted, not to recover the metals, but to produce compounds of sulphur. The roasting of pyrite produces sulphur dioxide gas, which can then be converted to sulphuric acid (see page 26).

Sulphur from fossil fuels
Although natural gas has a lower sulphur content than the other fossil fuels, it is still important to recover the sulphur to prevent pollution.

The sulphur is obtained by reacting hydrogen sulphide contained in the natural gas with oxygen. This produces sulphur dioxide gas, which can then be made into sulphuric acid by reacting the sulphur dioxide with oxygen and water.

Sulphur as a byproduct

All fossil fuels, as well as many metal ores, produce large volumes of sulphur-containing gas when they are heated. For a long time natural gas has been the main source of non-Frasch sulphur, because the sulphur had to be removed before the gas could be sold. Other producers of sulphur gases simply released them to the air.

With the tightening of regulations for the emission of sulphur, much of the sulphur required for modern industry is now recovered from fossil fuels by "scrubbing" the smokestack gases as they rise up from power stations and metal refineries that burn copper and lead sulphide ores.

Not only does this collect sulphur efficiently, but it reduces the amount of acid rain in the atmosphere. The extra production of sulphur from these sources has coincided with an increase in demand for sulphur for making fertilisers.

▲ Sulphur is dried and stored before use.

▶ The Frasch process of sulphur extraction involves pumping superheated water (water under pressure and heated to above 100°C) into a sulphur-bearing deposit. The sulphur does not react with water, so when it is pumped up to the surface the result is a suspension of sulphur in water. Another word for suspension is mixture.

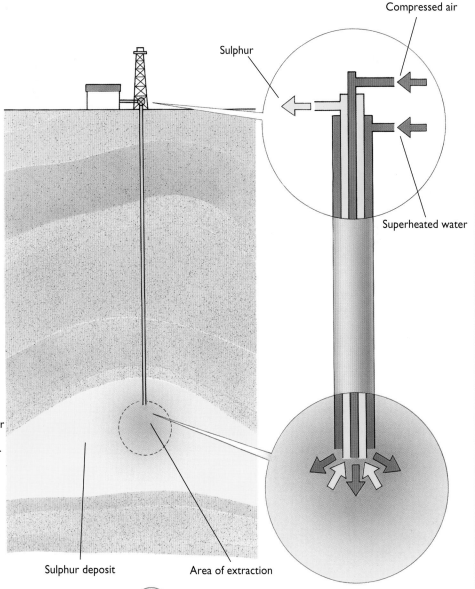

Compressed air

Sulphur

Superheated water

Sulphur deposit

Area of extraction

Sulphur dioxide

Sulphur dioxide is one of the most important sulphur gases. It is a dense, colourless gas with a choking (pungent) smell.

In large concentrations, sulphur dioxide can cause suffocation, while even small amounts can combine with water droplets to make acid rain. For these reasons, sulphur dioxide is regarded as a health hazard. In addition it is partly responsible for harming vegetation and corroding limestone buildings.

Sulphur dioxide is a pollutant, but also has many valuable uses. For example, it is a very good bleaching agent and many factories producing paper and textiles rely on it. Sulphur dioxide is also used in food warehouses because it is helpful in preserving both fruit and vegetables.

Preparing sulphur dioxide in the laboratory

Sulphur dioxide can be prepared by burning sulphur in oxygen as shown on pages 14 and 15, but more usually it is made by adding dilute sulphuric acid to a salt of sulphurous acid called a sulphite. Sulphites are generally used as bleaching agents.

EQUATION: Preparation of sulphur dioxide gas from a sulphur salt

Sulphuric acid + sodium sulphite ⇨ sulphur dioxide + sodium sulphate + water

$$H_2SO_4(aq) \quad + \quad Na_2SO_3(s) \quad ⇨ \quad SO_2(g) \quad + \quad Na_2SO_4(aq) \quad + \quad H_2O(l)$$

Properties of sulphur dioxide gas

Sulphur dioxide is an acidic gas, that is, when dissolved in water a weak acid is produced, called sulphurous acid.

Sulphur dioxide is a reducing agent, and will take oxygen from other substances. As a result, it can act as a bleach (see page 20).

Making use of sulphur dioxide

Sulphur dioxide can be used to produce sulphuric acid, a chemical used in fertilizer production. Some of the sulphur dioxide created by power stations is now recovered from the smokestacks rather than released into the environment (see page 22).

Sulphur dioxide can also be used as a bleaching agent because of its reducing properties. It is used to bleach wood pulp in making paper.

◄ Acid rain is responsible for damage to trees on already acid soils.

acid: compounds containing hydrogen which can attack and dissolve many substances. Acids are described as weak or strong, dilute or concentrated, mineral or organic.

strong acid: an acid that has completely dissociated (ionised) in water. Mineral acids are strong acids.

weak acid: an acid that has only partly dissociated (ionised) in water. Most organic acids are weak acids.

EQUATIONS: The reactions in the creation of acid rain
Stage 1: Sulphur dioxide gas emissions created by burning sulphur-containing impurities in petroleum

Hydrogen sulphide + oxygen ➪ water + sulphur dioxide

$$2H_2S(g) \quad + \quad 3O_2(g) \quad ➪ \quad 2H_2O(l) \quad + \quad 2SO_2(g)$$

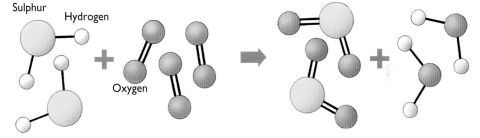

Sulphur
Hydrogen
Oxygen

How sulphur dioxide in air makes acid rain

When sulphur dioxide dissolves in water the result is sulphurous acid. This is part of the environmental problem called acid rain (see page 22). But sulphur dioxide is also oxidised by oxygen in the air to make sulphur trioxide. When sulphur trioxide dissolves in raindrops it produces the strong acid, sulphuric acid, which has a far more severe impact on the environment than the weak sulphurous acid.

Stage 2: Creating sulphur trioxide

Sulphur dioxide + oxygen ➪ sulphur trioxide

$$2SO_2(g) \quad + \quad O_2(g) \quad ➪ \quad 2SO_3(g)$$

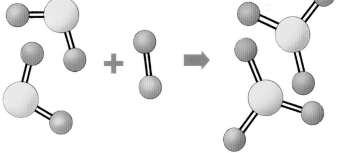

Stage 3: Dissolving sulphur trioxide in raindrops

Sulphur trioxide + water ➪ sulphuric acid

$$SO_3(g) \quad + \quad H_2O(l) \quad ➪ \quad H_2SO_4(aq)$$

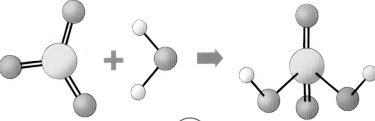

Sulphur dioxide as a bleaching agent

Any chemical that makes a material brighter and makes stains invisible is called a bleach. Bleaches also kill bacteria.

A bleach removes stains by reacting with the stain, making it colourless. Bleaches work either by adding oxygen to or removing it from the stained material. The colourless product is often more soluble and so is also more easily removed by detergents.

A number of sulphur-based products are used on a large scale for industrial bleaching. Sodium peroxydisulphate is used as a domestic bleach for use in home laundry applications because it is safer than chlorine-based bleaches on delicate textiles. It is an example of a bleach that oxidises the staining material.

Sodium sulphite and sulphur dioxide are used as industrial bleaches, especially for wood pulp. They act in the opposite way, reducing materials by removing oxygen from them.

Bread bleach
Sulphur dioxide is not a strong bleach and so can be used to bleach silk, wool and even flour (white flour for white bread). Sulphur dioxide will work on most dyes containing oxygen. In this case the sulphur dioxide acts as a reducing agent taking oxygen from the dye and making it colourless. But note that this process is reversible. The dye gradually picks up oxygen from the air and becomes coloured again. This is why, for example, straw hats, that were white when new, gradually turn yellow (the original colour of the straw containing natural dyes).

▲ White bread is made with flour that has been bleached using sulphur dioxide. Some brown bread also uses bleached white flour that has subsequently been dyed.

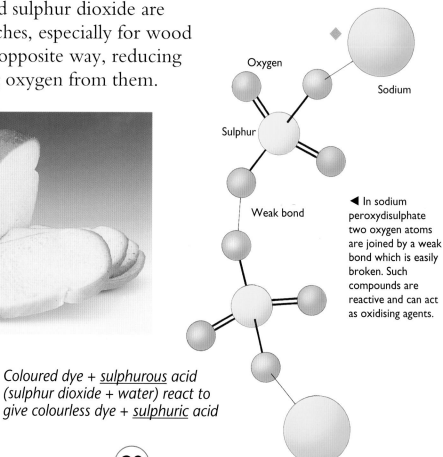

Oxygen

Sodium

Sulphur

Weak bond

◀ In sodium peroxydisulphate two oxygen atoms are joined by a weak bond which is easily broken. Such compounds are reactive and can act as oxidising agents.

Coloured dye + <u>sulphurous</u> acid (sulphur dioxide + water) react to give colourless dye + <u>sulphuric</u> acid

oxidising agent: a substance that removes electrons from another substance (and therefore is itself reduced).

reducing agent: a substance that gives electrons to another substance. Sulphur dioxide is a reducing agent, used for bleaching bread.

▲▶ These pansies show the effect of sulphur dioxide bleaching. A purple pansy was exposed to a sulphur dioxide atmosphere for two minutes and then put back in water, where it turned yellow. If it had been subjected to an even longer exposure it would have turned completely white (colourless).

Sulphur dioxide and the environment

The oxides of sulphur (sulphur dioxide, sulphur trioxide) are often referred to as "SO_x" for short. Sulphur dioxide is produced naturally when volcanoes erupt and when forests catch fire or vegetation decays.

But in our industrial world about as much sulphur dioxide is produced by industry (some 150 million tonnes a year) as from natural processes. This means that, instead of a natural amount of 10 parts per billion of sulphur dioxide in the air, the modern world experiences 20 parts per billion. In cities, on calm, foggy days, this level can rise to over 100 parts per billion, well within the zone of danger to health.

Polluting sulphur dioxide is produced when sulphurous fossil fuels are burned, so the main contributors are coal and oil-fired power stations and domestic heating systems. Natural gas is increasingly being used as a power station fuel because it contains relatively little sulphur. The fraction of petroleum used for making petrol also has little sulphur, which means that vehicles are not major producers of sulphur-based gases.

Smoke and sulphur dioxide

Smoke is a mixture of both small solids and gases, one of the most important of which is sulphur dioxide. When sulphur dioxide is breathed in, the gas combines with the water in the mouth and throat, adding to the acidity of these areas and often causing irritation. In higher concentrations it can cause sore throats.

However, when smoky fumes are breathed in, some of the sulphur dioxide is condensed on the tiny smoke particles and in this form it can be breathed right down into the lungs before it combines with water and turns into an acid. Here it can cause breathing problems and (like smoking) can also be a cause of lung cancer.

Smoke particles and sulphur dioxide were present in high concentrations in the Great London Smog (smoky fog) of 1952, when nearly 4000 people died in five days as a result of the suffocating air. This famous episode led to new laws in most industrial countries designed to keep levels of sulphur dioxide in the air under control. For example, smoky coals can no longer be burned in most city hearths or used to fuel boilers. Instead "smokeless" fuels must be used.

Also...

One of the most important catalysts in the air is an exhaust gas of vehicles, nitrogen dioxide. Sulphur dioxide will not naturally react with oxygen in the air, but nitrogen dioxide combines readily with sulphur dioxide to make the sulphur trioxide that will combine with raindrops to make sulphuric acid, which then falls as acid rain.

During the reaction, the nitrogen dioxide gives up some oxygen, but it can immediately get this back from the surrounding air, so in the end, no nitrogen dioxide is used up at all.

▶ A power station using brown coal (lignite). The clouds coming from the cooling towers are harmless water vapour. The polluting materials come from the tall chimneys (smokestacks). Traditionally, such chimneys were built tall to disperse the sulphur dioxide coming from them. Now most such power stations have "scrubbers" built into the chimney system to remove the sulphur dioxide. As a result, what is emitted from chimneys is now mostly particles of soot and carbon dioxide gas.

acidity: a general term for the strength of an acid in a solution.

smoke: a mixture of both small solids and gases.

Electrostatic precipitators

Many new power stations are being built with electrostatic precipitators built into the chimney system. An electrostatic precipitator uses static electrical charges to cause sulphur particles to be precipitated from the flue gases. This leaves the exhaust cleaner and less liable to cause acid rain. The sulphur collected can be used for other purposes (see pages 18 and 19), thus offsetting the cost of the equipment.

◀ These are electrostatic precipitators fitted to a new 1000 megawatt power station in India to remove the sulphur from the exhaust gases. The electrostatic precipitators are in the foreground of the picture.

EQUATION: Removing sulphur dioxide from power station exhausts with sodium hydroxide

Sulphur dioxide + sodium hydroxide ⇨ sodium sulphate + water

$$SO_2(g) \quad + \quad 2NaOH(aq) \quad ⇨ \quad Na_2SO_3(s) \quad + \quad H_2O(l)$$

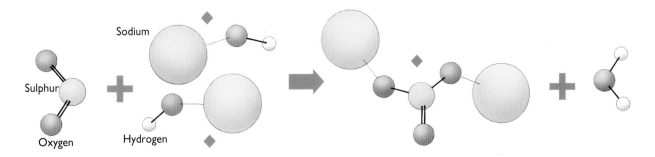

Sulphur
Oxygen
Sodium
Hydrogen

Pollution that crosses borders

Sulphur dioxide is a gas, so it can be carried large distances from where it is produced. This means that, for example, sulphur dioxide released by power stations in one place can be carried thousands of kilometres downwind, where it might combine with water droplets in a cloud and cause acid rain that harms trees.

In the most extreme cases, pollution caused in one country can cause acid rain damage in another country. This is most likely to be the case in both Europe and North America. In Europe the coals used in power stations in eastern Europe have produced pollution that has killed trees in the Black Forest of Germany; pollution produced in Britain may have harmed forests in Norway and Sweden; and pollution from the northeastern United States and southeastern Canada has harmed forests in eastern North America.

Hydrogen sulphide

Hydrogen sulphide gas burns with a blue flame. It is a reducing agent, taking oxygen from some of the compounds with which it reacts. It is poorly soluble in water, but when dissolved will produce a solution of hydrogen sulphide with a smell of bad eggs.

When hydrogen sulphide reacts with other compounds, the new sulphide compounds are characteristically black in colour (the exceptions are pyrite, which is yellow, and cadmium sulphide, which is orange). When passed over a solution of a lead salt, hydrogen sulphide will produce a precipitate of black lead sulphide. This is a laboratory test for hydrogen sulphide.

▲ Lead nitrate and hydrogen sulphide produce a black precipitate of lead sulphide (the mineral form of this is galena).

▼ Cobalt sulphide.

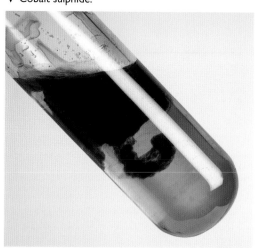

Preparing hydrogen sulphide in the laboratory
Hydrogen sulphide is normally prepared by adding dilute hydrochloric acid to iron sulphide.

EQUATION: Preparing hydrogen sulphide

Dilute hydrochloric acid + iron sulphide ⇨ ferric chloride + hydrogen sulphide

$$2HCl(aq) \quad + \quad FeS(s) \quad ⇨ \quad FeCl_2(aq) \quad + \quad H_2S(g)$$

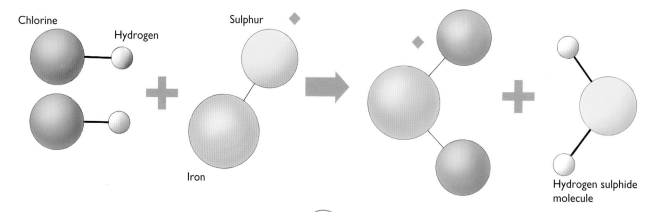

Chlorine

Hydrogen

Sulphur

Iron

Hydrogen sulphide molecule

▲ Nickel sulphide.

▼ Cadmium sulphide.

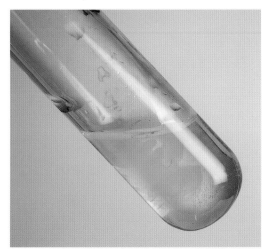

Hydrogen sulphide smells like rotten eggs

Hydrogen sulphide is present in natural gas and is also produced during the decay of dead material in the absence of water. This is one reason that some decaying remains and stagnant waters have a bad smell.

Hydrogen sulphide smells like rotten eggs. The odour occurs when the proteins of dead tissue that contain sulphur begin to decompose.

Tarnishing

Hydrogen sulphide is the polluting gas that reacts with silver in the atmosphere to produce black tarnishing (silver sulphide) on silverware.

Poisonous hydrogen sulphide

Hydrogen sulphide is a highly poisonous gas, which, because it has no colour, can only be detected by its smell. In fact, all the gases of sulphur (SO_X) are poisonous and most have unpleasant, choking odours.

Because SO_X gases quickly paralyse the nerves in the nose, their smell soon *appears* to go away. This is very dangerous effect because it encourages people to remain in contact with the gas, rather than move away.

Also...
Many fossil fuels contain sulphides, which are soluble in water. As water seeps into mine workings, it dissolves the sulphides and may subsequently find its way into groundwater or streams, where it can cause widespread and serious pollution. This is why the water pumped from active mines has to be carefully treated before being discharged.

EQUATION: Removing sulphur-containing impurities such as hydrogen sulphide from petroleum with sodium hydroxide

Hydrogen sulphide + sodium hydroxide ⇨ sodium sulphide + water

$$H_2S(g) + 2NaOH(aq) \Rightarrow Na_2S(s) + 2H_2O(l)$$

Sulphuric acid

Sulphuric acid, a colourless, thick, heavy and "oily" liquid, is one of the most common acids used in industry and the laboratory. It is a strong mineral acid.

Dilute sulphuric acid will react with most metals to form sulphate compounds and release hydrogen gas. It reacts with bases to form sulphates and water only.

Concentrated sulphuric acid is also a dehydrating agent, that is, it will remove water from other compounds. For example, if a piece of sugar is placed in sulphuric acid it becomes a black lump of carbon, as all the water is removed from the sugar molecules.

If sulphuric acid touches human skin, the molecules in the skin immediately begin to lose water. This is an acid burn.

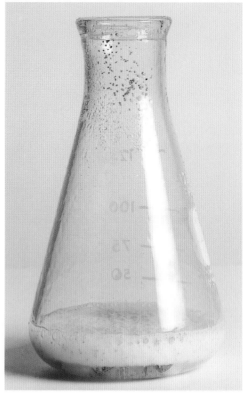

▲ Sulphuric acid reacts with metals such as zinc to produce a sulphate of the metal and release hydrogen gas.

Also...

When water and a concentrated acid mix, a huge amount of heat is given off. Water should never be added to concentrated acid. Rather, the acid should be stirred slowly into the water. If water is added to the acid, the water will immediately start to boil causing it to splash out much like hot fat in a fryer. The result could be dangerous acid burns.

EQUATION: Reaction of dilute sulphuric acid and zinc metal

Sulphuric acid + zinc ⇨ zinc sulphate + hydrogen

$$H_2SO_4(aq) \quad + \quad Zn(s) \quad ⇨ \quad ZnSO_4(aq) \quad + \quad H_2(g)$$

A dehydrating agent

The demonstration on these two pages shows how water can be removed from sugar by dehydration using sulphuric acid.

❶▶ A layer of sugar is placed in a beaker and concentrated sulphuric acid is added.

EQUATION: Dehydration of sucrose using concentrated sulphuric acid

Sucrose + sulphuric acid ⇨ steam + carbon + sulphuric acid

$$C_{12}H_{22}O_{11}(s) \quad + \quad H_2SO_4(l) \quad ⇨ \quad 11H_2O(g) \quad + \quad 12C(s) \quad + \quad H_2SO_4(aq)$$

corrosive: a substance, either an acid or an alkali, that *rapidly* attacks a wide range of other substances.

mineral acid: an acid that does not contain carbon and that attacks minerals. Hydrochloric, sulphuric and nitric acids are the main mineral acids.

strong acid: an acid that has completely dissociated (ionised) in water. Mineral acids are strong acids.

❷▲❸▼ The sugar will turn yellow, brown and finally black.

❹▲ Suddenly the surface will bulge up and crack. By now so much heat is given out that much of the surplus water produced by the reaction will form steam, but the steam bubbles are trapped in the sticky carbon and they simply expand in the "goo", making it frothy. When it sets it is like a piece of coke.

Manufacturing sulphuric acid

More sulphuric acid is made than almost any other substance. This is because the acid is useful in making so many other products.

Sulphuric acid is made by reacting sulphur dioxide with oxygen and water. Sulphur dioxide is obtained either from hydrogen sulphide in natural gas, or as pure sulphur is burned in air. It may also be collected from power station smokestacks.

Sulphur dioxide reacts only slowly with oxygen, so the process is made quicker by reacting the substances at a high temperature (500°C) and in the presence of a catalyst. The catalyst used in this reaction (called the Contact process) is a material called vanadium pentoxide.

Sulphur dioxide and air are forced through a tower containing the catalyst in the form of pellets. The sulphur trioxide gas then flows into a tower containing quartz bathed in a circulating bath of sulphuric acid. As the gas dissolves in the sulphuric acid, the acid becomes more concentrated and some is drawn off and diluted.

The gas is not added directly to water because this would give off too much heat energy and make the water evaporate.

The Contact process

Sulphur dioxide and oxygen are brought together in a converter containing vanadium pentoxide. Usually more than one stage of conversion is needed. The reaction produces heat. If the reacting gases get too hot they stop reacting, so the gases are led from one converter, cooled and then fed into another. Up to four stages of this conversion are needed. By this time nearly all of the sulphur dioxide has been converted to sulphur trioxide.

The absorber is a tank of sulphuric acid that absorbs sulphur trioxide, concentrating the sulphuric acid even further. Some of this liquid can be drawn off and diluted as required. Sulphur trioxide is not added to water because it releases a great deal of heat and produces a fine mist that is difficult to handle.

◀ A bank of Contact process converters.

▼ The Contact process.

catalyst: a substance that speeds up a chemical reaction but itself remains unaltered at the end of the reaction.

Sulphur dioxide (SO_2) and oxygen (O_2)

① Converter containing vanadium pentoxide catalyst achieves 63% conversion to sulphur trioxide.

Gases cooled

② Converter containing vanadium pentoxide catalyst achieves 84% conversion to sulphur trioxide.

Gases cooled

③ Converter containing vanadium pentoxide catalyst achieves 93% conversion to sulphur trioxide.

SO_3 fed to an absorber

Remaining SO_3

Sulphur trioxide absorbed to create 99.5% sulphuric acid, which is then diluted to 98% acid.

④ Converter containing vanadium pentoxide catalyst achieves 99.5% conversion to sulphur trioxide.

Gases fed to an absorber

Remaining sulphur trioxide absorbed to create 99.5% sulphuric acid, which is then diluted to 98% acid.

EQUATION: Stage 1: Reacting sulphur dioxide and oxygen to produce sulphur trioxide

Sulphur dioxide gas + oxygen ⇨ sulphur trioxide

$$2SO_2(g) \quad + \quad O_2(g) \quad ⇨ \quad 2SO_3(g)$$

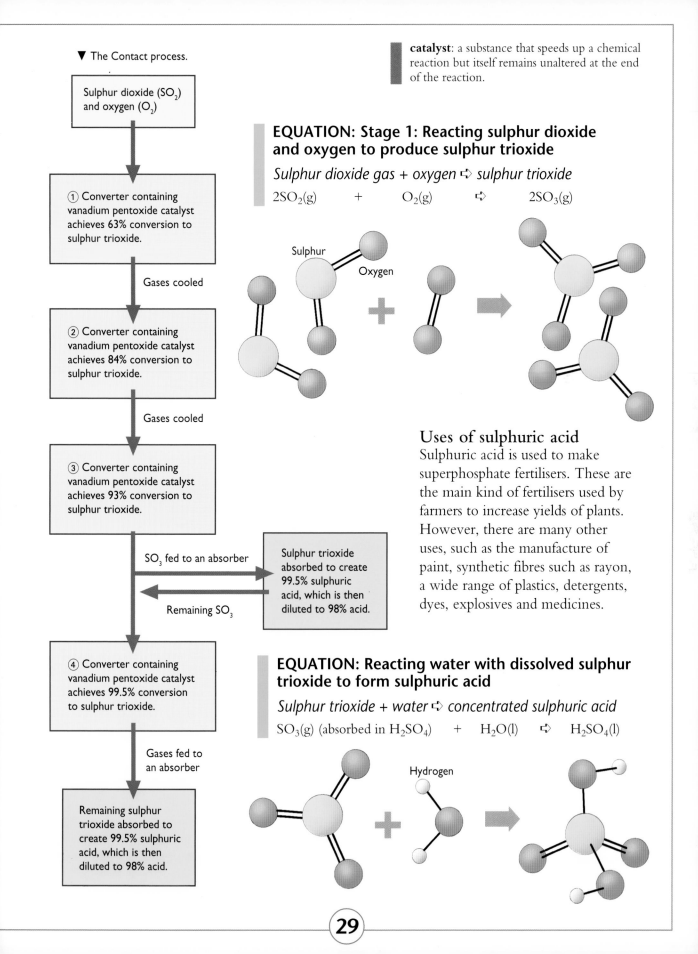

Sulphur

Oxygen

Uses of sulphuric acid

Sulphuric acid is used to make superphosphate fertilisers. These are the main kind of fertilisers used by farmers to increase yields of plants. However, there are many other uses, such as the manufacture of paint, synthetic fibres such as rayon, a wide range of plastics, detergents, dyes, explosives and medicines.

EQUATION: Reacting water with dissolved sulphur trioxide to form sulphuric acid

Sulphur trioxide + water ⇨ concentrated sulphuric acid

$$SO_3(g) \text{ (absorbed in } H_2SO_4) \quad + \quad H_2O(l) \quad ⇨ \quad H_2SO_4(l)$$

Hydrogen

Sulphuric acid as an electrolyte

Dilute sulphuric acid is used in vehicle batteries, sometimes called "wet batteries", and more accurately called "secondary batteries". They are called secondary batteries because the cells inside the battery can be recharged. (Contrast this with a primary cell, like a dry cell, where the chemicals inside the cell react to produce electricity only once). In a vehicle battery six cells are combined to give the twelve volt supply that vehicles use worldwide.

Lead–acid batteries are the most commonly used form of vehicle battery. To work as an electrical battery, each of the cells inside it must have two electrodes made of electrically conducting materials. These must be bathed in a liquid that can conduct electricity as well as help the battery store electricity. This liquid is called an electrolyte.

▶ Lead-acid batteries are designed to give each electrode a large surface area. This allows it to change stored chemical energy into electrical energy quickly. The electrodes are made from a lead alloy; half are covered in a paste of lead sulphate. All the electrodes are bathed in dilute sulphuric acid.

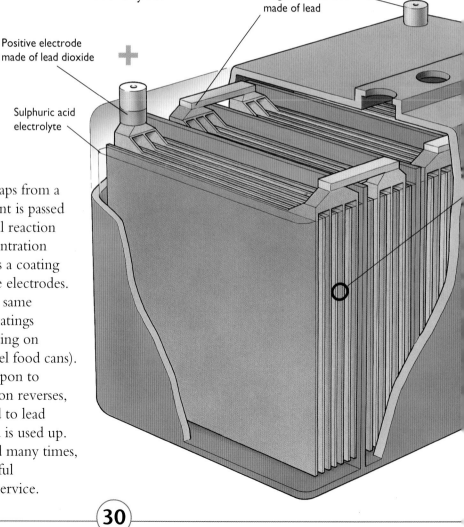

Negative electrode made of lead

Positive electrode made of lead dioxide

Sulphuric acid electrolyte

Operation of the battery

As the battery is charged (perhaps from a vehicle generator), and a current is passed through the battery, a chemical reaction occurs that increases the concentration of the sulphuric acid and forms a coating of lead and lead dioxide on the electrodes. This process is electrolysis (the same process that puts chromium coatings on cutlery, gold and silver-plating on ornaments and tin-plate on steel food cans).

When the battery is called upon to discharge electricity, the reaction reverses, the lead coatings are converted to lead sulphate and the sulphuric acid is used up.

This process can be repeated many times, giving the secondary cell a useful life of many years of constant service.

▼ A secondary battery works by converting chemical energy to electrical energy. For this to happen there have to be two different materials as electrodes, and an electron-carrying liquid, an electrolyte. The condition of the electrolyte changes as it charges and discharges.

acidity: a general term for the strength of an acid in a solution.

electrode: a conductor that forms one terminal of a cell.

electrolyte: a solution that conducts electricity.

ion: an atom, or group of atoms, that has gained or lost one or more electrons and so developed an electrical charge. Ions carry electrical current through solutions.

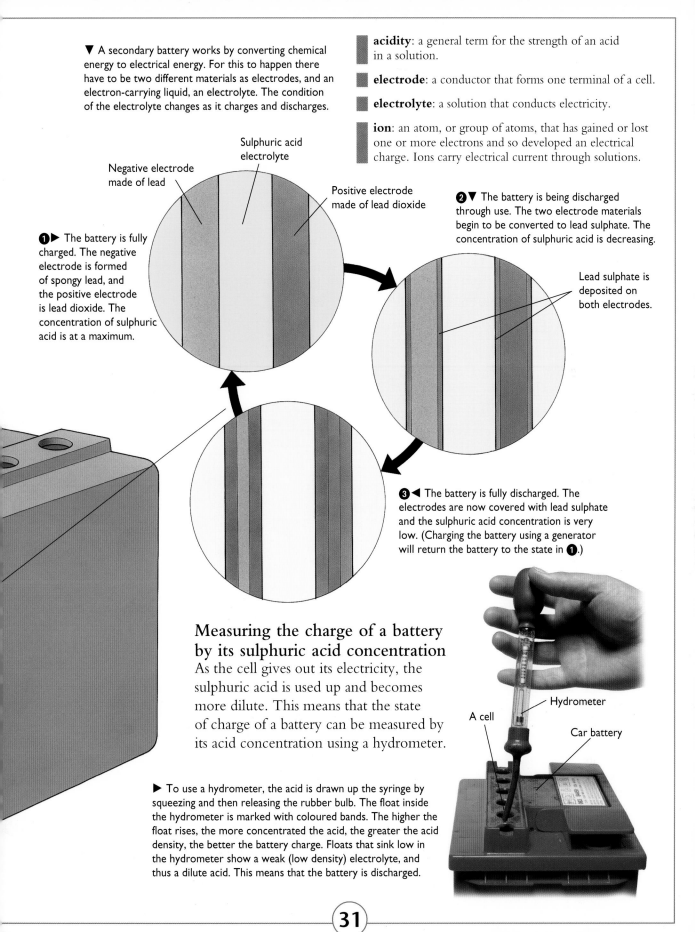

Sulphuric acid electrolyte

Negative electrode made of lead

Positive electrode made of lead dioxide

❶▶ The battery is fully charged. The negative electrode is formed of spongy lead, and the positive electrode is lead dioxide. The concentration of sulphuric acid is at a maximum.

❷▼ The battery is being discharged through use. The two electrode materials begin to be converted to lead sulphate. The concentration of sulphuric acid is decreasing.

Lead sulphate is deposited on both electrodes.

❸◀ The battery is fully discharged. The electrodes are now covered with lead sulphate and the sulphuric acid concentration is very low. (Charging the battery using a generator will return the battery to the state in ❶.)

Measuring the charge of a battery by its sulphuric acid concentration

As the cell gives out its electricity, the sulphuric acid is used up and becomes more dilute. This means that the state of charge of a battery can be measured by its acid concentration using a hydrometer.

Hydrometer

A cell

Car battery

▶ To use a hydrometer, the acid is drawn up the syringe by squeezing and then releasing the rubber bulb. The float inside the hydrometer is marked with coloured bands. The higher the float rises, the more concentrated the acid, the greater the acid density, the better the battery charge. Floats that sink low in the hydrometer show a weak (low density) electrolyte, and thus a dilute acid. This means that the battery is discharged.

Sulphates and sulphites

Sulphates are compounds containing sulphur and oxygen. Sulphites are compounds of sulphur containing less oxygen than sulphates.

There is a wide range of sulphates, usually formed by a combination of a metal (such as calcium, magnesium or sodium, as shown on this page, or copper as shown on page 34).

Sulphate salts are formed in the laboratory by reacting sulphuric acid on a metal, oxide, hydroxide or carbonate.

Most sulphates are soluble in water.

▲ Gypsum is used for plastering in the construction industry.

Magnesium sulphate

Magnesium sulphate was first made by evaporating mineral water from a spring near a town called Epsom in Surrey, England. For this reason it is also called Epsom salts.

Epsom salts act as a laxative by preventing the water in the intestine from being absorbed in the body. The result is that there is more water than usual in the intestine, which sets the action of the bowels in motion.

Concentrated Epsom salts are used to stop inflammation and to prevent convulsions. They are also used in both dyeing and bleaching processes, as part of fertilizers, in matches and explosives. They can also have a fireproofing effect.

Calcium sulphate

One of the most common sulphate minerals in the Earth's rocks is calcium sulphate, also known as gypsum. One form of gypsum is alabaster which, being soft, is easily machined to make ornaments such as candlesticks.

Gypsum is formed in the bottom of saline lakes. As the lake water evaporates, the chemicals dissolved in it become progressively more concentrated until the solution becomes saturated and crystals begin to form.

Crushed gypsum is used as Plaster of Paris and is made into wallboards (it is the smooth surface of the interiors of walls).

▼ Sulphates and their properties

Sulphate	Common name	Application
Calcium sulphate	Gypsum Plaster of Paris	Soil conditioner, plasterboard Casts (used to set broken limbs)
Barium sulphate	Barite	Drilling mud in oil fields, digestion, X–rays
Sodium sulphate	Glauber's salt	Glass-making
Lead sulphate		Produced as lead–acid battery discharges
Copper sulphate	Blue vitriol	Fungicide (Bordeaux mixture)
Iron sulphate	Green vitriol	Lawn care
Magnesium sulphate	Epsom salts	Laxative

Sodium sulphate

Also known as Glauber's salt, sodium sulphate is commonly found dissolved in drinking water. Sodium sulphate occurs as thick evaporite beds and can be mined by pumping hot water down a well and pumping the dissolved sulphate back up.

Sodium sulphate is used instead of sulphuric acid for some processes, for example in making dyes, paper and glass, and to help molten metals flow (when used as a flux).

Sodium sulphate crystals melt at 32°C, meaning that they will melt at high room temperatures. On melting they absorb very large amounts of heat (the salt has a large latent heat). This means that sodium sulphate can be used in some forms of solar storage heating. The Sun's rays can be used to melt the sulphate, and later, as it cools and solidifies during the night, it releases its heat.

▲ Sodium thiosulphate crystals

flux: a material used to make it easier for a liquid to flow. A flux dissolves metal oxides and so prevents a metal from oxidising while being heated.

latent heat: the amount of heat that is absorbed or released during the process of changing state between gas, liquid or solid. For example, heat is absorbed when a substance melts and it is released again when the substance solidifies.

saturated solution: a solution that holds the maximum possible amount of dissolved material. The amount of material in solution varies with the temperature; cold solutions can hold less dissolved solid material than hot solutions. Gases are more soluble in cold liquids than hot liquids.

Also...Photographer's hypo

Every time you have a photograph printed, the printer will have used a chemical called hypo. This is short for the chemical sodium thiosulphate. It is used to "fix" the image after it has been developed.

The photographic film contains a surface coating, or emulsion, of chemicals. This emulsion, a kind of gelatin, contains minute crystals or grains of silver compounds spread evenly over it The smaller the size of the crystals, the finer the grain and the better the eventual picture quality. (This is why we talk about "graininess" when referring to photograph quality.)

When the camera shutter is worked, the iris opens on the camera and light reaches the emulsion of the film. The light causes a chemical reaction to occur that results in small groups of silver atoms clumping together. Wherever silver atoms clump together within a crystal or grain they will provide the image on the film.

Developing a negative requires several chemical stages. The first chemical stage happens in a darkroom, where the developer solution converts the clumps of silver atoms into tiny particles of pure silver.

However, the emulsion still contains a mix of chemicals that would react to light and fog the film; this mix has to be removed before the film can be exposed to the light. This is the job of the hypo. The hypo dissolves away the remaining silver-containing compounds. After fixing, the film is stable, and all other processing can be done in normal light.

The hypo reacts with the silver compounds of the emulsion that have not been changed to pure silver, but leaves the pure silver alone. When the film is washed in water, everything except the silver is removed and the film surface contains just silver. The film is now a "negative" in which all light areas are shown dark and all dark areas are shown transparent or light. The positive print (the actual photograph) is produced by shining a light through the negative onto a piece of light-sensitive printing paper.

EQUATION: Making photographer's hypo

Sodium sulphite + sulphur ⇨ sodium thiosulphate

$$Na_2SO_3(aq) \quad + \quad S(s) \quad \Rightarrow \quad Na_2S_2O_3(aq)$$

Copper sulphate

Copper sulphate occurs as a blue solution or crystals. It can be made by reacting sulphuric acid and copper carbonate.

Copper sulphate is widely used as a fungicide. Originally it was probably used in a pure form as a seed covering to protect the seed from rotting before it could germinate. Later it was combined with lime to make Bordeaux mixture (see page 42). It is also used to treat wood and preserve it from attack by fungi, moulds and other rotting organisms.

A white form of copper sulphate (called anhydrous copper sulphate) will absorb large amounts of water and can therefore be used as a drying agent. Anhydrous copper sulphate can also be used as a simple test for the presence of water because it changes from white to blue when in contact with water.

Preparing copper sulphate

Dilute sulphuric acid and green copper carbonate produce a blue solution of copper sulphate.

▲► Sulphuric acid is added to copper carbonate powder to produce a solution of copper sulphate that can be evaporated, leaving copper sulphate crystals. A copper sulphate solution can also be produced by adding sulphuric acid to black copper oxide powder.

EQUATION: Reaction of sulphuric acid and copper carbonate

Sulphuric acid + copper carbonate ⇨ copper sulphate + carbon dioxide + water

$$H_2SO_4(aq) \quad + \quad CuCO_3(s) \quad ⇨ \quad CuSO_4(aq) \quad + \quad CO_2(g) \quad + \quad H_2O(l)$$

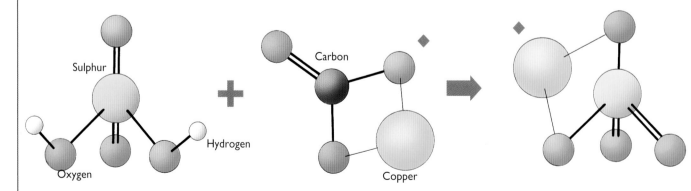

Sulphur

Hydrogen

Oxygen

Carbon

Copper

Sulphuric acid   *Copper carbonate*   *Copper sulphate*

anhydrous: a substance from which water has been removed by heating. Many hydrated salts are crystalline. When they are heated and the water is driven off, the material changes to an anhydrous powder.

hydrate: a solid compound in crystalline form that contains molecular water. Hydrates commonly form when a solution of a soluble salt is evaporated. The water that forms part of a hydrate crystal is known as the "water of crystallization". It can usually be removed by heating, leaving an anhydrous salt.

Hydrated copper sulphate

▲▼ Hydrated copper sulphate (shown above) has been left next to a heat source. The water has been driven off to give anhydrous copper sulphate (shown below).

Anhydrous copper sulphate

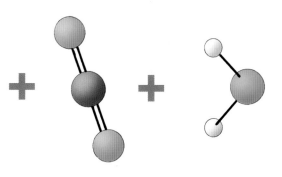

Carbon dioxide   Water

35

Reactions with copper sulphate

Copper sulphate is widely used in laboratory demonstrations. The colour makes the reactions easier to see. Here are some typical reactions.

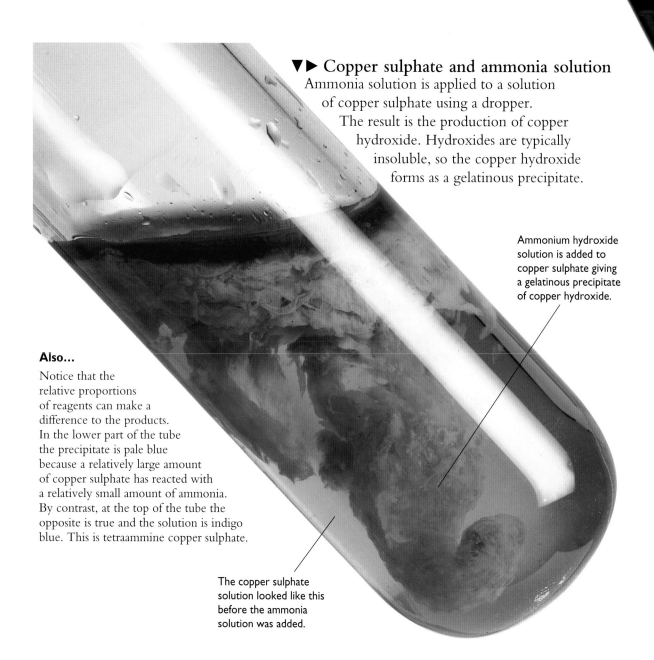

▼▶ Copper sulphate and ammonia solution
Ammonia solution is applied to a solution of copper sulphate using a dropper. The result is the production of copper hydroxide. Hydroxides are typically insoluble, so the copper hydroxide forms as a gelatinous precipitate.

Ammonium hydroxide solution is added to copper sulphate giving a gelatinous precipitate of copper hydroxide.

Also...
Notice that the relative proportions of reagents can make a difference to the products. In the lower part of the tube the precipitate is pale blue because a relatively large amount of copper sulphate has reacted with a relatively small amount of ammonia. By contrast, at the top of the tube the opposite is true and the solution is indigo blue. This is tetraammine copper sulphate.

The copper sulphate solution looked like this before the ammonia solution was added.

EQUATION: Copper sulphate and ammonia

Copper sulphate + ammonium solution ⇨ copper hydroxide + ammonium sulphate

$CuSO_4(aq)$ + $2NH_4OH(aq)$ ⇨ $Cu(OH)_2(aq)$ + $(NH_4)_2SO_4(aq)$

The gelatinous precipitate of copper hydroxide is redissolved as more concentrated ammonia solution is added to produce a dark blue copper compound called a "complex".

gelatinous: a term meaning made with water. Because a gelatinous precipitate is mostly water, it is of a similar density to water and will float or lie suspended in the liquid.

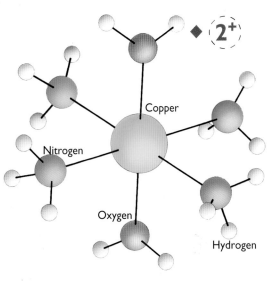

2^+

Copper

Nitrogen

Oxygen

Hydrogen

▲ A diagrammatic representation of the structure of the copper complex formed when excess ammonia solution is added to copper sulphate.

▶ Copper sulphate and hydrogen sulphide

A hydrogen sulphide solution is added to copper sulphate from a dropper. The result is a precipitate of copper sulphide. Compare this precipitate to the copper hydroxide produced in the demonstration on the opposite page.

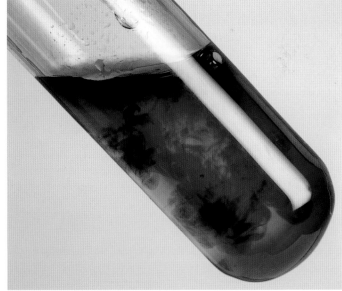

EQUATION: Copper sulphate and hydrogen sulphide

Copper sulphate + hydrogen sulphide ⇨ copper sulphide + sulphuric acid

$CuSO_4(aq)$ + $H_2S(aq)$ ⇨ $CuS(s)$ + $H_2SO_4(aq)$

Vulcanising rubber

Vulcanising is the addition of sulphur to rubber. American inventor Charles Goodyear discovered the vulcanising effect of sulphur by chance in 1839 when trying to find an improved form of rubber.

Rubber is a polymer, that is a material made of long chains of subunits called monomers. The rubber is soft and elastic because the chains of atoms move relatively freely across each other. In the process of heating rubber with sulphur, the sulphur atoms bond with the chains of rubber, linking one chain to another. This makes it harder for the chains of rubber to move against each other and the result is a stronger material.

Vulcanised rubber does not soften as it is heated in the way that natural rubber does and so can be used for tyres and other applications where durability and strength is important.

An extreme form of vulcanising causes so many sulphur atoms to link together the rubber chains that the material becomes a solid. This solid is called ebonite.

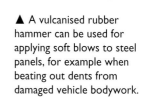

▲ A vulcanised rubber hammer can be used for applying soft blows to steel panels, for example when beating out dents from damaged vehicle bodywork.

Also...

Sulphur atoms are found in human fingernails, where they crosslink the proteins to form a rigid material. So when you look at your nails, you are looking at a natural form of vulcanising.

monomer: a building block of a larger chain molecule ("mono" means one, "mer" means part).

polymer: a compound that is made of long chains by combining molecules (called monomers) as repeating units. ("Poly" means many, "mer" means part).

polymerisation: a chemical reaction in which large numbers of similar molecules arrange themselves into large molecules, usually long chains. This process usually happens when there is a suitable catalyst present. For example, ethene reacts to form polythene in the presence of certain catalysts.

vulcanisation: forming cross–links between polymer chains to increase the strength of the whole polymer. Rubbers are vulcanised using sulphur when making tyres and other strong materials.

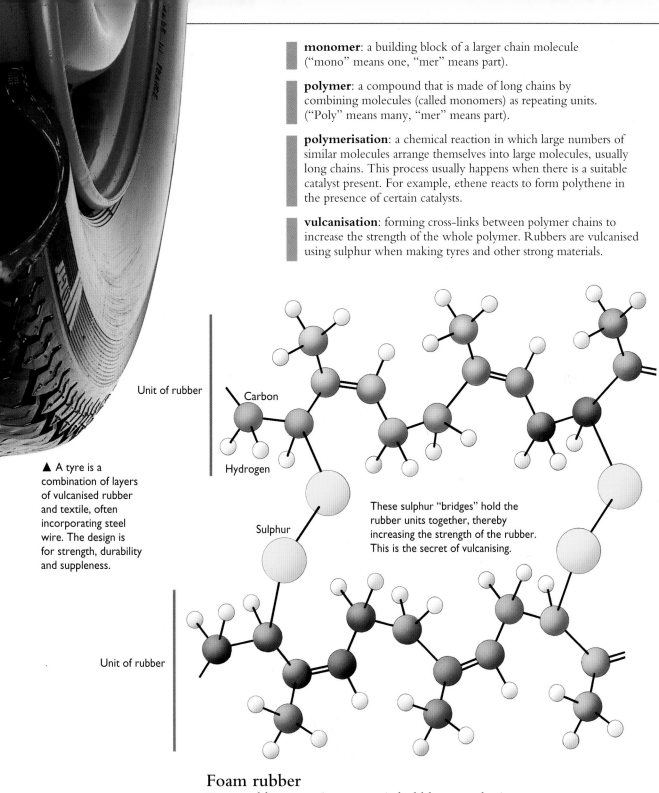

▲ A tyre is a combination of layers of vulcanised rubber and textile, often incorporating steel wire. The design is for strength, durability and suppleness.

Unit of rubber

Carbon

Hydrogen

Sulphur

Unit of rubber

These sulphur "bridges" hold the rubber units together, thereby increasing the strength of the rubber. This is the secret of vulcanising.

Foam rubber

Foam rubber contains many air bubbles to make it spongy and lightweight. Its main use is for fillings in household furniture, mattresses and carpet backing.

The latex is frothed up to incorporate air and then vulcanised in such a way that water is not evaporated from the foam before the vulcanisation process is complete.

Sulphur in warfare

Sulphur can be put to many uses. Among them are the manufacture of gunpowder and poison gas.

Gunpowder is a solid mixture of sulphur, charcoal and potassium nitrate that was invented in China about one thousand years ago. It is the earliest form of explosive and is still widely used.

Sulphur reacts with chlorine to produce a yellow liquid with a revolting smell that is used in the vulcanisation of rubber. This liquid has peaceful uses and is not poisonous. However, by adding more chlorine a foul-smelling red liquid is produced. If this is in turn reacted with ethene, it produces a gas that not only has a choking smell, but is extremely poisonous. This poison gas is called mustard gas. It was used in World War I, and its effects were so horrific that it was banned by international agreement. Nevertheless, in recent years some countries have used it again, often in internal conflicts.

Mustard gas

Mustard gas (dichlorodiethyl sulphide) is a poison gas that, unfortunately, is quite easy to make and use. It is made as an oily liquid, which slowly evaporates when released into the environment.

When mustard gas is breathed into the lungs it damages the cells on the lung lining. This causes fluid to leave the blood and fill the lungs. As a result, a person suffering from mustard gas poisoning drowns.

In less severe amounts it causes permanent lung damage and blisters on the skin. In its most characteristic form it causes blindness.

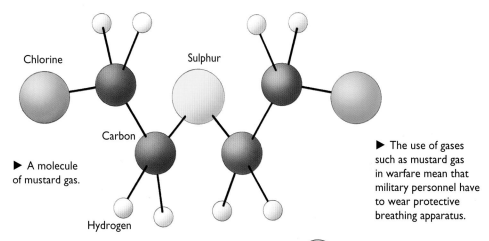

Chlorine

Sulphur

Carbon

▶ A molecule of mustard gas.

Hydrogen

▶ The use of gases such as mustard gas in warfare mean that military personnel have to wear protective breathing apparatus.

explosive: a substance which, when a shock is applied to it, decomposes very rapidly, releasing a very large amount of heat and creating a large volume of gas as a shock wave.

high explosive: a form of explosive that will only work when it receives a shock from another explosive. High explosives are much more powerful than ordinary explosives. Gunpowder is not a high explosive.

poison gas: a form of gas that is used intentionally to produce widespread injury and death. (Many gases are poisonous, which is why many chemical reactions are performed in laboratory fume chambers, but they are a byproduct of a reaction and not intended to cause harm.)

Gunpowder

Gunpowder is the best known and earliest form of explosive. It is made by mixing powders of sulphur, charcoal and saltpetre (potassium nitrate) together. The first people to invent gunpowder were the Chinese, and gunpowder has been in use for nearly one thousand years.

Gunpowder is ignited by means of a spark (say from tinder in a flintlock rifle), a flame (light the fuse and stand clear, as in fireworks) or an electric arc (as in detonation with an electrical plunger).

When gunpowder was first made, it was mixed and then crushed into a powder by people using hammers. As a result, from time to time, the shock of the hammers caused the powder to explode. Making gunpowder was never a good trade to be in!

The modern manufacture of gunpowder – now called black powder – is by pulverising charcoal with sulphur and then mixing it with potassium nitrate so that the particles of nitrate, sulphur and charcoal all come into close contact. To do this, very heavy steel wheels are rolled over a black powder mixture spread on a steel plate. The mixture is then pressed into a cake and broken down into granules.

The main modern use of gunpowder is in fireworks. For this purpose the gunpowder is mixed with graphite, which coats the powder and makes it less likely to explode as it is carried about. Elsewhere, black powder is used as a first charge to set off artillery shells and other munitions.

Sulphur for life

Sulphur originally came from gases in volcanic eruptions. It is now an essential nutrient in almost all living things. Much of the sulphur in living things is recycled between generations, a process called the sulphur cycle.

Sulphur-based medicines were among the first to deal with microbial and bacterial infections and are still widely used.

The poisonous properties of some sulphur compounds are also used to help prevent infections and to preserve foods.

Preservatives

Sulphites are used widely for food preservation. They have a wide range of actions. They may kill off fungi and bacteria, preventing decay; they may stop oxygen reacting with foodstuffs, another source of decay.

Sulphites may also be used to clean vessels in which food may be placed. This is the reason, for example, that sulphite is added to water and then used to sterilise beer and wine vessels both in factories and in amateur beer and wine making.

Sulphites are also able to stop the natural decay processes that cause food to become discoloured. For this reason their use has been restricted so that food cannot be stored and treated with sulphites to make it look fresher than it is.

Sulphur dioxide gas is also used as a preservative. Sulphur dioxide and sulphites both work by taking oxygen from the air, preventing microorganisms from getting the oxygen they need to live, and by creating a mildly acid environment in which organisms cannot survive.

▶ **Sulphur drugs**
Sulphur compounds have been used in the treatment of many illnesses since ancient times. For example, brimstone and treacle was a traditional "catch-all" remedy.

Modern sulphur medicines include antibacterial drugs such as sulphonamides. They work by destroying the enzyme that is needed for the growth of bacterial cells.

Sulphur and pesticides

A pest is any life-form that causes illness and discomfort to people, or affects their food supplies or gardens.

Pesticides (named from the Latin suffix *cide*, which means to kill) are chemicals that are designed to control these pests, whether they be plants or animals. From the writings of the ancient Greek, Homer, we can tell that as long ago as 1000 BC, ancient civilisations were aware of the pesticide properties of sulphur.

Both the Chinese and Greeks knew about the lethal combination of sulphur and arsenic. Another of the old pesticides was Bordeaux mixture, a mixture of copper sulphate and lime. It still is used to combat fungi and to repel insects.

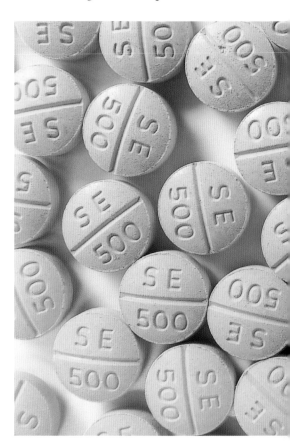

Sulphate fertilisers

Sulphates are soluble in water. This is important for plants because they need to take up their nutrients through their roots using water in the soil. Those nutrients most needed by plants to make their tissues are potassium, nitrogen and phosphorus. Sulphates are thus used as a cheap "vehicle" to carry sulphur and the other main nutrients that the plants need.

A fertiliser is produced by reacting a suitable source of potassium, nitrogen or phosphorus with sulphuric acid. About two-thirds of all the sulphuric acid made goes into the production of fertilisers.

Most fertilisers contain one or more of ammonium sulphate (nitrogen), calcium phosphate (which contains phosphorus), and sulphate of potash (potassium sulphate).

One of the best known fertilisers is called superphosphate (a combination of ammonium sulphate and phosphate), first produced by scientist John Lawes in 1843. To make it, sulphuric acid is poured over phosphate rock, producing phosphoric acid. Ammonia gas and sulphuric acid are also reacted to produce ammonium sulphate. The two fertilisers are then mixed to make superphosphate.

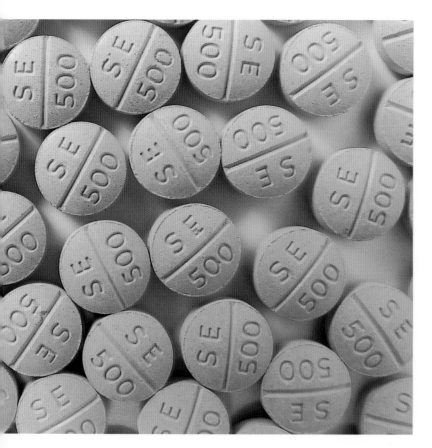

pesticide: any chemical that is designed to control pests (unwanted organisms) that are harmful to plants or animals.

preservative: a substance that prevents the natural organic decay processes from occurring. Many substances can be used safely for this purpose, including sulphites and nitrogen gas.

sulphate: a compound that includes sulphur and oxygen, for example, calcium sulphate or gypsum.

sulphide: a sulphur compound that contains no oxygen.

sulphite: a sulphur compound that contains less oxygen than a sulphate.

Energy for life

Many different organisms use sulphur instead of oxygen to oxidise their energy supplies. Among the most common are bacteria. This is why bacteria and blue–green algae are able to live in places with very little oxygen. The vastness of the deep ocean waters is one example; the areas near geysers and volcanoes are another.

Whereas green plants give off oxygen, organisms that live in oxygen-poor water give off sulphur as part of their living processes. These organisms get all the energy they need by fermenting dead tissue using sulphur-containing compounds.

Also...

In recent years farmers near industrial regions have noticed that their yields of cereals have declined almost in proportion to the rate at which sulphur has been removed from the smokestacks of power stations. This is because plants have been using the sulphur in the acid rain as a fertiliser. Now acid rain has been reduced, farmers are having to increase their applications of sulphate fertilisers in order to maintain their yields!

Key facts about...
Sulphur

Has no taste

Has no smell

Poor conductor of
electricity and heat

Can be found as native
sulphur in crystalline form

A very reactive element
that combines with
almost all other elements

A mustard-yellow solid,
chemical symbol S

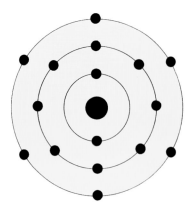

Melts at 119°C, just a
little over the boiling
point of water

Soft and can be scratched
with a fingernail

Density 2 g/ml, about
twice that of water

Insoluble in water

Atomic number 16,
atomic weight about 32

SHELL DIAGRAMS
The shell diagrams on this page
are representations of an atom of
each element. The total number
of electrons are shown in the
relevant orbitals, or shells, around
the central nucleus.

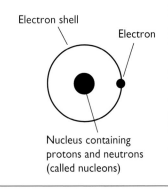

Electron shell

Electron

Nucleus containing
protons and neutrons
(called nucleons)

▶ A hydrogen sulphide solution is
added to silver nitrate, leaving a dark
brown precipitate of silver sulphide.
Sulphide precipitates are mostly a
dark colour and are frequently black.

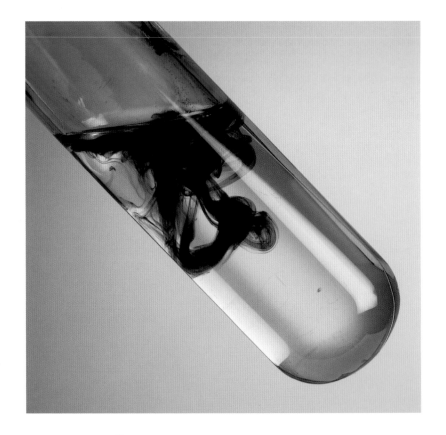

▼ "Crash-cooling" molten sulphur causes the atoms to form long chains and the sulphur forms a plastic-like substance that turns brittle as the atoms assume a more stable structure (see pages 6 and 7 for more about this demonstration).

The Periodic Table

The Periodic Table sets out the relationships among the elements of the Universe. According to the Periodic Table, certain elements fall into groups. The pattern of these groups has, in the past, allowed scientists to predict elements that had not at that time been discovered. It can still be used today to predict the properties of unfamiliar elements.

The Periodic Table was first described by a Russian teacher, Dmitry Ivanovich Mendeleev, between 1869 and 1870. He was interested in writing a chemistry textbook, and wanted to show his students that there were certain patterns in the elements that had been discovered. So he set out the elements (of which there were 57 at the time) according to their known properties. On the assumption that there was pattern to the elements, he left blank spaces where elements seemed to be missing. Using this first version of the Periodic Table, he was able to predict in detail the chemical and physical properties of elements that had not yet been discovered. Other scientists began to look for the missing elements, and they soon found them.

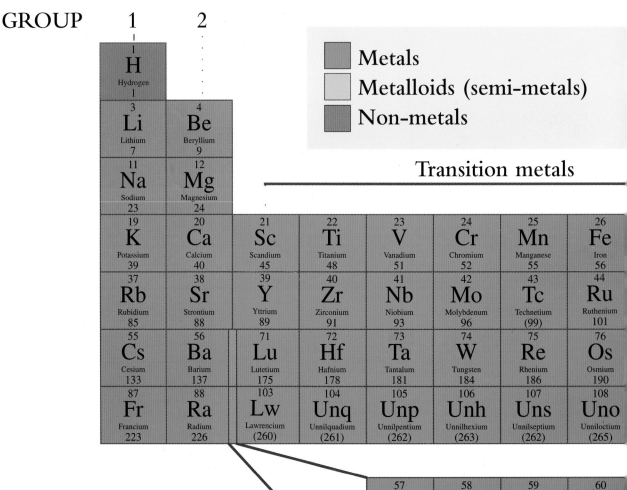

GROUP 1 2

Metals
Metalloids (semi-metals)
Non-metals

Transition metals

1 H Hydrogen 1							
3 Li Lithium 7	4 Be Beryllium 9						
11 Na Sodium 23	12 Mg Magnesium 24						
19 K Potassium 39	20 Ca Calcium 40	21 Sc Scandium 45	22 Ti Titanium 48	23 V Vanadium 51	24 Cr Chromium 52	25 Mn Manganese 55	26 Fe Iron 56
37 Rb Rubidium 85	38 Sr Strontium 88	39 Y Yttrium 89	40 Zr Zirconium 91	41 Nb Niobium 93	42 Mo Molybdenum 96	43 Tc Technetium (99)	44 Ru Ruthenium 101
55 Cs Cesium 133	56 Ba Barium 137	71 Lu Lutetium 175	72 Hf Hafnium 178	73 Ta Tantalum 181	74 W Tungsten 184	75 Re Rhenium 186	76 Os Osmium 190
87 Fr Francium 223	88 Ra Radium 226	103 Lw Lawrencium (260)	104 Unq Unnilquadium (261)	105 Unp Unnilpentium (262)	106 Unh Unnilhexium (263)	107 Uns Unnilseptium (262)	108 Uno Unniloctium (265)

Lanthanide metals

| 57 La Lanthanum 139 | 58 Ce Cerium 140 | 59 Pr Praseodymium 141 | 60 Nd Neodymium 144 |

Actinoid metals

| 89 Ac Actinium (227) | 90 Th Thorium 232 | 91 Pa Protactinium 231 | 92 U Uranium 238 |

Hydrogen did not seem to fit into the table, so he placed it in a box on its own. Otherwise the elements were all placed horizontally. When an element was reached with properties similar to the first one in the top row, a second row was started. By following this rule, similarities among the elements can be found by reading up and down. By reading across the rows, the elements progressively increase their atomic number. This number indicates the number of positively charged particles (protons) in the nucleus of each atom. This is also the number of negatively charged particles (electrons) in the atom.

The chemical properties of an element depend on the number of electrons in the outermost shell.

Atoms can form compounds by sharing electrons in their outermost shells. This explains why atoms with a full set of electrons (like helium, an inert gas) are unreactive, whereas atoms with an incomplete electron shell (such as chlorine) are very reactive. Elements can also combine by the complete transfer of electrons from metals to non-metals and the compounds formed contain ions.

Radioactive elements lose particles from their nucleus and electrons from their surrounding shells. As a result their atomic number changes and they become new elements.

Key:
- Atomic (proton) number — 13
- Symbol — Al
- Name — Aluminium
- Approximate relative atomic mass (Approximate atomic weight) — 27

3	4	5	6	7	0
					2 He Helium 4
5 B Boron 11	6 C Carbon 12	7 N Nitrogen 14	8 O Oxygen 16	9 F Fluorine 19	10 Ne Neon 20
13 Al Aluminium 27	14 Si Silicon 28	15 P Phosphorus 31	16 S Sulphur 32	17 Cl Chlorine 35	18 Ar Argon 40

				3	4	5	6	7	0
27 Co Cobalt 59	28 Ni Nickel 59	29 Cu Copper 64	30 Zn Zinc 65	31 Ga Gallium 70	32 Ge Germanium 73	33 As Arsenic 75	34 Se Selenium 79	35 Br Bromine 80	36 Kr Krypton 84
45 Rh Rhodium 103	46 Pd Palladium 106	47 Ag Silver 108	48 Cd Cadmium 112	49 In Indium 115	50 Sn Tin 119	51 Sb Antimony 122	52 Te Tellurium 128	53 I Iodine 127	54 Xe Xenon 131
77 Ir Iridium 192	78 Pt Platinum 195	79 Au Gold 197	80 Hg Mercury 201	81 Tl Thallium 204	82 Pb Lead 207	83 Bi Bismuth 209	84 Po Polonium (209)	85 At Astatine (210)	86 Rn Radon (222)
109 Une Unnilennium (266)									

61 Pm Promethium (145)	62 Sm Samarium 150	63 Eu Europium 152	64 Gd Gadolinium 157	65 Tb Terbium 159	66 Dy Dysprosium 163	67 Ho Holmium 165	68 Er Erbium 167	69 Tm Thulium 169	70 Yb Ytterbium 173
93 Np Neptunium (237)	94 Pu Plutonium (244)	95 Am Americium (243)	96 Cm Curium (247)	97 Bk Berkelium (247)	98 Cf Californium (251)	99 Es Einsteinium (252)	100 Fm Fermium (257)	101 Md Mendelevium (258)	102 No Nobelium (259)

Understanding equations

As you read through this book, you will notice that many pages contain equations using symbols. If you are not familiar with these symbols, read this page. Symbols make it easy for chemists to write out the reactions that are occurring in a way that allows a better understanding of the processes involved.

Symbols for the elements

The basis of the modern use of symbols for elements dates back to the 19th century. At this time a shorthand was developed using the first letter of the element wherever possible. Thus "O" stands for oxygen, "H" stands for hydrogen

and so on. However, if we were to use only the first letter, then there could be some confusion. For example, nitrogen and nickel would both use the symbols N. To overcome this problem, many elements are symbolised using the first two letters of their full name, and the second letter is lowercase. Thus although nitrogen is N, nickel becomes Ni. Not all symbols come from the English name; many use the Latin name instead. This is why, for example, gold is not G but Au (for the Latin *aurum*) and sodium has the symbol Na, from the Latin *natrium*.

Compounds of elements are made by combining letters. Thus the molecule carbon

Written and symbolic equations

In this book, important chemical equations are briefly stated in words (these are called word equations), and are then shown in their symbolic form along with the states.

What reaction the equation illustrates

EQUATION: The formation of calcium hydroxide

Word equation

Calcium oxide + water ⇨ calcium hydroxide

Symbol equation

$$CaO(s) \quad + \quad H_2O(l) \quad \overset{\Rightarrow}{\text{heated}} \quad Ca(OH)_2(aq)$$

Sometimes you will find additional descriptions below the symbolic equation.

Symbol showing the state: *s* is for solid, *l* is for liquid, *g* is for gas and *aq* is for aqueous.

Diagrams

Some of the equations are shown as graphic representations.

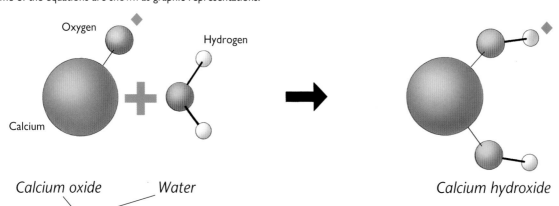

Oxygen

Hydrogen

Calcium

Calcium oxide   *Water*

Calcium hydroxide

Sometimes the written equation is broken up and put below the relevant stages in the graphic representation.

monoxide is CO. By using lowercase letters for the second letter of an element, it is possible to show that cobalt, symbol Co, is not the same as the molecule carbon monoxide, CO.

However, the letters can be made to do much more than this. In many molecules, atoms combine in unequal numbers. So, for example, carbon dioxide has one atom of carbon for every two of oxygen. This is shown by using the number 2 beside the oxygen, and the symbol becomes CO_2.

In practice, some groups of atoms combine as a unit with other substances. Thus, for example, calcium bicarbonate (one of the compounds used in some antacid pills) is written $Ca(HCO_3)_2$. This shows that the part of the substance inside the brackets reacts as a unit and the "2" outside the brackets shows the presence of two such units.

Some substances attract water molecules to themselves. To show this a dot is used. Thus the blue form of copper sulphate is written $CuSO_4.5H_2O$. In this case five molecules of water attract to one of copper sulphate.

When you see the dot, you know that this water can be driven off by heating; it is part of the crystal structure.

In a reaction substances change by rearranging the combinations of atoms. The way they change is shown by using the chemical symbols, placing those that will react (the starting materials, or reactants) on the left and the products of the reaction on the right. Between the two, chemists use an arrow to show which way the reaction is occurring.

It is possible to describe a reaction in words. This gives word equations, which are given throughout this book. However, it is easier to understand what is happening by using an equation containing symbols. These are also given in many places. They are not given when the equations are very complex.

In any equation both sides balance; that is, there must be an equal number of like atoms on both sides of the arrow. When you try to write down reactions, you, too, must balance your equation; you cannot have a few atoms left over at the end!

The symbols in brackets are abbreviations for the physical state of each substance taking part, so that (s) is used for solid, (l) for liquid, (g) for gas and (aq) for an aqueous solution, that is, a solution of a substance dissolved in water.

Atoms and ions
Each sphere represents a particle of an element. A particle can be an atom or an ion. Each atom or ion is associated with other atoms or ions through bonds – forces of attraction. The size of the particles and the nature of the bonds can be extremely important in determining the nature of the reaction or the properties of the compound.

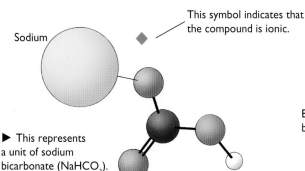

Sodium

This symbol indicates that the compound is ionic.

▶ This represents a unit of sodium bicarbonate ($NaHCO_3$).

The term "unit" is sometimes used to simplify the representation of a combination of ions.

Chemical symbols, equations and diagrams
The arrangement of any molecule or compound can be shown in one of the two ways shown below, depending on which gives the clearer picture. The left-hand diagram is called a ball-and-stick diagram because it uses rods and spheres to show the structure of the material. This example shows water, H_2O. There are two hydrogen atoms and one oxygen atom.

Bond shown by "stick"

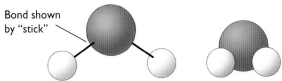

Colours too
The colours of each of the particles help differentiate the elements involved. The diagram can then be matched to the written and symbolic equation given with the diagram. In the case above, oxygen is red and hydrogen is grey.

Glossary of technical terms

absorb: to soak up a substance. Compare to adsorb.

acetone: a petroleum-based solvent.

acid: compounds containing hydrogen which can attack and dissolve many substances. Acids are described as weak or strong, dilute or concentrated, mineral or organic.

acidity: a general term for the strength of an acid in a solution.

acid rain: rain that is contaminated by acid gases such as sulphur dioxide and nitrogen oxides released by pollution.

adsorb/adsorption: to "collect" gas molecules or other particles on to the *surface* of a substance. They are not chemically combined and can be removed. (The process is called "adsorption".) Compare to absorb.

alchemy: the traditional "art" of working with chemicals that prevailed through the Middle Ages. One of the main challenges of alchemy was to make gold from lead. Alchemy faded away as scientific chemistry was developed in the 17th century.

alkali: a base in solution.

alkaline: the opposite of acidic. Alkalis are bases that dissolve, and alkaline materials are called basic materials. Solutions of alkalis have a pH greater than 7.0 because they contain relatively few hydrogen ions.

alloy: a mixture of a metal and various other elements.

alpha particle: a stable combination of two protons and two neutrons, which is ejected from the nucleus of a radioactive atom as it decays. An alpha particle is also the nucleus of the atom of helium. If it captures two electrons it can become a neutral helium atom.

amalgam: a liquid alloy of mercury with another metal.

amino acid: amino acids are organic compounds that are the building blocks for the proteins in the body.

amorphous: a solid in which the atoms are not arranged regularly (i.e. "glassy"). Compare with crystalline.

amphoteric: a metal that will react with both acids and alkalis.

anhydrous: a substance from which water has been removed by heating. Many hydrated salts are crystalline. When they are heated and the water is driven off, the material changes to an anhydrous powder.

anion: a negatively charged atom or group of atoms.

anode: the negative terminal of a battery or the positive electrode of an electrolysis cell.

anodising: a process that uses the effect of electrolysis to make a surface corrosion-resistant.

antacid: a common name for any compound that reacts with stomach acid to neutralise it.

antioxidant: a substance that prevents oxidation of some other substance.

aqueous: a solid dissolved in water. Usually used as "aqueous solution".

atom: the smallest particle of an element.

atomic number: the number of electrons or the number of protons in an atom.

atomised: broken up into a very fine mist. The term is used in connection with sprays and engine fuel systems.

aurora: the "northern lights" and "southern lights" that show as coloured bands of light in the night sky at high latitudes. They are associated with the way cosmic rays interact with oxygen and nitrogen in the air.

basalt: an igneous rock with a low proportion of silica (usually below 55%). It has microscopically small crystals.

base: a compound that may be soapy to the touch and that can react with an acid in water to form a salt and water.

battery: a series of electrochemical cells.

bauxite: an ore of aluminium, of which about half is aluminium oxide.

becquerel: a unit of radiation equal to one nuclear disintegration per second.

beta particle: a form of radiation in which electrons are emitted from an atom as the nucleus breaks down.

bleach: a substance that removes stains from materials either by oxidising or reducing the staining compound.

boiling point: the temperature at which a liquid boils, changing from a liquid to a gas.

bond: chemical bonding is either a transfer or sharing of electrons by two or more atoms. There are a number of types of chemical bond, some very strong (such as covalent bonds), others weak (such as hydrogen bonds). Chemical bonds form because the linked molecule is more stable than the unlinked atoms from which it formed. For example, the hydrogen molecule (H_2) is more stable than single atoms of hydrogen, which is why hydrogen gas is always found as molecules of two hydrogen atoms.

brass: a metal alloy principally of copper and zinc.

brazing: a form of soldering, in which brass is used as the joining metal.

brine: a solution of salt (sodium chloride) in water.

bronze: an alloy principally of copper and tin.

buffer: a chemistry term meaning a mixture of substances in solution that resists a change in the acidity or alkalinity of the solution.

capillary action: the tendency of a liquid to be sucked into small spaces, such as between objects and through narrow-pore tubes. The force to do this comes from surface tension.

catalyst: a substance that speeds up a chemical reaction but itself remains unaltered at the end of the reaction.

cathode: the positive terminal of a battery or the negative electrode of an electrolysis cell.

cathodic protection: the technique of making the object that is to be protected from corrosion into the cathode of a cell. For example, a material, such as steel, is protected by coupling it with a more reactive metal, such as magnesium. Steel forms the cathode and magnesium the anode. Zinc protects steel in the same way.

cation: a positively charged atom or group of atoms.

caustic: a substance that can cause burns if it touches the skin.

cell: a vessel containing two electrodes and an electrolyte that can act as an electrical conductor.

ceramic: a material based on clay minerals, which has been heated so that it has chemically hardened.

chalk: a pure form of calcium carbonate made of the crushed bodies of microscopic sea creatures, such as plankton and algae.

change of state: a change between one of the three states of matter, solid, liquid and gas.

chlorination: adding chlorine to a substance.

cladding: a surface sheet of material designed to protect other materials from corrosion.

clay: a microscopically small plate-like mineral that makes up the bulk of many soils. It has a sticky feel when wet.

combustion: the special case of oxidisation of a substance where a considerable amount of heat and usually light are given out. Combustion is often referred to as "burning".

compound: a chemical consisting of two or more elements chemically bonded together. Calcium atoms can combine with carbon atoms and oxygen atoms to make calcium carbonate, a compound of all three atoms.

condensation nuclei: microscopic particles of dust, salt and other materials suspended in the air, which attract water molecules.

conduction: (i) the exchange of heat (heat conduction) by contact with another object or (ii) allowing the flow of electrons (electrical conduction).

convection: the exchange of heat energy with the surroundings produced by the flow of a fluid due to being heated or cooled.

corrosion: the *slow* decay of a substance resulting from contact with gases and liquids in the environment. The term is often applied to metals. Rust is the corrosion of iron.

corrosive: a substance, either an acid or an alkali, that *rapidly* attacks a wide range of other substances.

cosmic rays: particles that fly through space and bombard all atoms on the Earth's surface. When they interact with the atmosphere they produce showers of secondary particles.

covalent bond: the most common form of strong chemical bonding, which occurs when two atoms *share* electrons.

cracking: breaking down complex molecules into simpler components. It is a term particularly used in oil refining.

crude oil: a chemical mixture of petroleum liquids. Crude oil forms the raw material for an oil refinery.

crystal: a substance that has grown freely so that it can develop external faces. Compare with crystalline, where the atoms are not free to form individual crystals and amorphous where the atoms are arranged irregularly.

crystalline: the organisation of atoms into a rigid "honeycomb-like" pattern without distinct crystal faces.

crystal systems: seven patterns or systems into which all of the world's crystals can be grouped. They are: cubic, hexagonal, rhombohedral, tetragonal, orthorhombic, monoclinic and triclinic.

cubic crystal system: groupings of crystals that look like cubes.

curie: a unit of radiation. The amount of radiation emitted by 1 g of radium each second. (The curie is equal to 37 billion becquerels.)

current: an electric current is produced by a flow of electrons through a conducting solid or ions through a conducting liquid.

decay (radioactive decay): the way that a radioactive element changes into another element because of loss of mass through radiation. For example uranium decays (changes) to lead.

decompose: to break down a substance (for example by heat or with the aid of a catalyst) into simpler components. In such a chemical reaction only one substance is involved.

dehydration: the removal of water from a substance by heating it, placing it in a dry atmosphere, or through the action of a drying agent.

density: the mass per unit volume (e.g. g/cc).

desertification: a process whereby a soil is allowed to become degraded to a state in which crops can no longer grow, i.e. desert-like. Chemical desertification is usually the result of contamination with halides because of poor irrigation practices.

detergent: a petroleum-based chemical that removes dirt.

diaphragm: a semipermeable membrane – a kind of ultra-fine mesh filter – that will allow only small ions to pass through. It is used in the electrolysis of brine.

diffusion: the slow mixing of one substance with another until the two substances are evenly mixed.

digestive tract: the system of the body that forms the pathway for food and its waste products. It begins at the mouth and includes the stomach and the intestines.

dilute acid: an acid whose concentration has been reduced by a large proportion of water.

diode: a semiconducting device that allows an electric current to flow in only one direction.

disinfectant: a chemical that kills bacteria and other microorganisms.

dissociate: to break apart. In the case of acids it means to break up forming hydrogen ions. This is an example of ionisation. Strong acids dissociate completely. Weak acids are not completely ionised and a solution of a weak acid has a relatively low concentration of hydrogen ions.

dissolve: to break down a substance in a solution without a resultant reaction.

distillation: the process of separating mixtures by condensing the vapours through cooling.

doping: adding metal atoms to a region of silicon to make it semiconducting.

dye: a coloured substance that will stick to another substance, so that both appear coloured.

electrode: a conductor that forms one terminal of a cell.

electrolysis: an electrical–chemical process that uses an electric current to cause the break up of a compound and the movement of metal ions in a solution. The process happens in many natural situations (as for example in rusting) and is also commonly used in industry for purifying (refining) metals or for plating metal objects with a fine, even metal coating.

electrolyte: a solution that conducts electricity.

electron: a tiny, negatively charged particle that is part of an atom. The flow of electrons through a solid material such as a wire produces an electric current.

electroplating: depositing a thin layer of a metal onto the surface of another substance using electrolysis.

element: a substance that cannot be decomposed into simpler substances by chemical means

emulsion: tiny droplets of one substance dispersed in another. A common oil in water emulsion is milk. The tiny droplets in an emulsion tend to come together, so another stabilising substance is often needed to wrap the particles of grease and oil in a stable coat. Soaps and detergents are such agents. Photographic film is an example of a solid emulsion.

endothermic reaction: a reaction that takes heat from the surroundings. The reaction of carbon monoxide with a metal oxide is an example.

enzyme: organic catalysts in the form of proteins in the body that speed up chemical reactions. Every living cell contains hundreds of enzymes, which ensure that the processes of life continue. Should enzymes be made inoperative, such as through mercury poisoning, then death follows.

ester: organic compounds, formed by the reaction of an alcohol with an acid, which often have a fruity taste.

evaporation: the change of state of a liquid to a gas. Evaporation happens below the boiling point and is used as a method of separating out the materials in a solution.

exothermic reaction: a reaction that gives heat to the surroundings. Many oxidation reactions, for example, give out heat.

explosive: a substance which, when a shock is applied to it, decomposes very rapidly, releasing a very large amount of heat and creating a large volume of gases as a shock wave.

extrusion: forming a shape by pushing it through a die. For example, toothpaste is extruded through the cap (die) of the toothpaste tube.

fallout: radioactive particles that reach the ground from radioactive materials in the atmosphere.

fat: semi-solid energy-rich compounds derived from plants or animals and which are made of carbon, hydrogen and oxygen. Scientists call these esters.

feldspar: a mineral consisting of sheets of aluminium silicate. This is the mineral from which the clay in soils is made.

fertile: able to provide the nutrients needed for unrestricted plant growth.

filtration: the separation of a liquid from a solid using a membrane with small holes.

fission: the breakdown of the structure of an atom, popularly called "splitting the atom" because the atom is split into approximately two other nuclei. This is different from, for example, the small change that happens when radioactivity is emitted.

fixation of nitrogen: the processes that natural organisms, such as bacteria, use to turn the nitrogen of the air into ammonium compounds.

fixing: making solid and liquid nitrogen-containing compounds from nitrogen gas. The compounds that are formed can be used as fertilisers.

fluid: able to flow; either a liquid or a gas.

fluorescent: a substance that gives out visible light when struck by invisible waves such as ultraviolet rays.

flux: a material used to make it easier for a liquid to flow. A flux dissolves metal oxides and so prevents a metal from oxidising while being heated.

foam: a substance that is sufficiently gelatinous to be able to contain bubbles of gas. The gas bulks up the substance, making it behave as though it were semi-rigid.

fossil fuels: hydrocarbon compounds that have been formed from buried plant and animal remains. High pressures and temperatures lasting over millions of years are required. The fossil fuels are coal, oil and natural gas.

fraction: a group of similar components of a mixture. In the petroleum industry the light fractions of crude oil are those with the smallest molecules, while the medium and heavy fractions have larger molecules.

free radical: a very reactive atom or group with a "spare" electron.

freezing point: the temperature at which a substance changes from a liquid to a solid. It is the same temperature as the melting point.

fuel: a concentrated form of chemical energy. The main sources of fuels (called fossil fuels because they were formed by geological processes) are coal, crude oil and natural gas. Products include methane, propane and gasoline. The fuel for stars and space vehicles is hydrogen.

fuel rods: rods of uranium or other radioactive material used as a fuel in nuclear power stations.

fuming: an unstable liquid that gives off a gas. Very concentrated acid solutions are often fuming solutions.

fungicide: any chemical that is designed to kill fungi and control the spread of fungal spores.

fusion: combining atoms to form a heavier atom.

galvanising: applying a thin zinc coating to protect another metal.

gamma rays: waves of radiation produced as the nucleus of a radioactive element rearranges itself into a tighter cluster of protons and neutrons. Gamma rays carry enough energy to damage living cells.

gangue: the unwanted material in an ore.

gas: a form of matter in which the molecules form no definite shape and are free to move about to fill any vessel they are put in.

gelatinous: a term meaning made with water. Because a gelatinous precipitate is mostly water, it is of a similar density to water and will float or lie suspended in the liquid.

gelling agent: a semi-solid jelly-like substance.

gemstone: a wide range of minerals valued by people, both as crystals (such as emerald) and as decorative stones (such as agate). There is no single chemical formula for a gemstone.

glass: a transparent silicate without any crystal growth. It has a glassy lustre and breaks with a curved fracture. Note that some minerals have all these features and are therefore natural glasses. Household glass is a synthetic silicate.

glucose: the most common of the natural sugars. It occurs as the polymer known as cellulose, the fibre in plants. Starch is also a form of glucose. The breakdown of glucose provides the energy that animals need for life.

granite: an igneous rock with a high proportion of silica (usually over 65%). It has well-developed large crystals. The largest pink, grey or white crystals are feldspar.

Greenhouse Effect: an increase of the global air temperature as a result of heat released from burning fossil fuels being absorbed by carbon dioxide in the atmosphere.

gypsum: the name for calcium sulphate. It is commonly found as Plaster of Paris and wallboards.

half-life: the time it takes for the radiation coming from a sample of a radioactive element to decrease by half.

halide: a salt of one of the halogens (fluorine, chlorine, bromine and iodine).

halite: the mineral made of sodium chloride.

halogen: one of a group of elements including chlorine, bromine, iodine and fluorine.

heat-producing: see exothermic reaction.

high explosive: a form of explosive that will only work when it receives a shock from another explosive. High explosives are much more powerful than ordinary explosives. Gunpowder is not a high explosive.

hydrate: a solid compound in crystalline form that contains molecular water. Hydrates commonly form when a solution of a soluble salt is evaporated. The water that forms part of a hydrate crystal is known as the "water of crystallization". It can usually be removed by heating, leaving an anhydrous salt.

hydration: the absorption of water by a substance. Hydrated materials are not "wet" but remain firm, apparently dry, solids. In some cases, hydration makes the substance change colour, in many other cases there is no colour change, simply a change in volume.

hydrocarbon: a compound in which only hydrogen and carbon atoms are present. Most fuels are hydrocarbons, as is the simple plastic polyethene (known as polythene).

hydrogen bond: a type of attractive force that holds one molecule to another. It is one of the weaker forms of intermolecular attractive force.

hydrothermal: a process in which hot water is involved. It is usually used in the context of rock formation because hot water and other fluids sent outwards from liquid magmas are important carriers of metals and the minerals that form gemstones.

igneous rock: a rock that has solidified from molten rock, either volcanic lava on the Earth's surface or magma deep underground. In either case the rock develops a network of interlocking crystals.

incendiary: a substance designed to cause burning.

indicator: a substance or mixture of substances that change colour with acidity or alkalinity.

inert: nonreactive.

infra-red radiation: a form of light radiation where the wavelength of the waves is slightly longer than visible light. Most heat radiation is in the infra-red band.

insoluble: a substance that will not dissolve.

ion: an atom, or group of atoms, that has gained or lost one or more electrons and so developed an electrical charge. Ions behave differently from electrically neutral atoms and molecules. They can move in an electric field, and they can also bind strongly to solvent molecules such as water. Positively charged ions are called cations; negatively charged ions are called anions. Ions carry electrical current through solutions.

ionic bond: the form of bonding that occurs between two ions when the ions have opposite charges. Sodium cations bond with chloride anions to form common salt (NaCl) when a salty solution is evaporated. Ionic bonds are strong bonds except in the presence of a solvent.

ionise: to break up neutral molecules into oppositely charged ions or to convert atoms into ions by the loss of electrons.

ionisation: a process that creates ions.

irrigation: the application of water to fields to help plants grow during times when natural rainfall is sparse.

isotope: atoms that have the same number of protons in their nucleus, but which have different masses; for example, carbon-12 and carbon-14.

latent heat: the amount of heat that is absorbed or released during the process of changing state between gas, liquid or solid. For example, heat is absorbed when a substance melts and it is released again when the substance solidifies.

latex: (the Latin word for "liquid") a suspension of small polymer particles in water. The rubber that flows from a rubber tree is a natural latex. Some synthetic polymers are made as latexes, allowing polymerisation to take place in water.

lava: the material that flows from a volcano.

limestone: a form of calcium carbonate rock that is often formed of lime mud. Most limestones are light grey and have abundant fossils.

liquid: a form of matter that has a fixed volume but no fixed shape.

lode: a deposit in which a number of veins of a metal found close together.

lustre: the shininess of a substance.

magma: the molten rock that forms a balloon-shaped chamber in the rock below a volcano. It is fed by rock moving upwards from below the crust.

marble: a form of limestone that has been "baked" while deep inside mountains. This has caused the limestone to melt and reform into small interlocking crystals, making marble harder than limestone.

mass: the amount of matter in an object. In everyday use, the word weight is often used to mean mass.

melting point: the temperature at which a substance changes state from a solid to a liquid. It is the same as freezing point.

membrane: a thin flexible sheet. A semipermeable membrane has microscopic holes of a size that will selectively allow some ions and molecules to pass through but hold others back. It thus acts as a kind of sieve.

meniscus: the curved surface of a liquid that forms when it rises in a small bore, or capillary tube. The meniscus is convex (bulges upwards) for mercury and is concave (sags downwards) for water.

metal: a substance with a lustre, the ability to conduct heat and electricity and which is not brittle.

metallic bonding: a kind of bonding in which atoms reside in a "sea" of mobile electrons. This type of bonding allows metals to be good conductors and means that they are not brittle.

metamorphic rock: formed either from igneous or sedimentary rocks, by heat and or pressure. Metamorphic rocks form deep inside mountains during periods of mountain building. They result from the remelting of rocks during which process crystals are able to grow. Metamorphic rocks often show signs of banding and partial melting.

micronutrient: an element that the body requires in small amounts. Another term is trace element.

mineral: a solid substance made of just one element or chemical compound. Calcite is a mineral because it consists only of calcium carbonate, halite is a mineral because it contains only sodium chloride, quartz is a mineral because it consists of only silicon dioxide.

mineral acid: an acid that does not contain carbon and that attacks minerals. Hydrochloric, sulphuric and nitric acids are the main mineral acids.

mineral-laden: a solution close to saturation.

mixture: a material that can be separated out into two or more substances using physical means.

molecule: a group of two or more atoms held together by chemical bonds.

monoclinic system: a grouping of crystals that look like double-ended chisel blades.

monomer: a building block of a larger chain molecule ("mono" means one, "mer" means part).

mordant: any chemical that allows dyes to stick to other substances.

native metal: a pure form of a metal, not combined as a compound. Native metal is more common in poorly reactive elements than in those that are very reactive.

neutralisation: the reaction of acids and bases to produce a salt and water. The reaction causes hydrogen from the acid and hydroxide from the base to be changed to water. For example, hydrochloric acid reacts with sodium hydroxide to form common salt and water. The term is more generally used for any reaction where the pH changes towards 7.0, which is the pH of a neutral solution.

neutron: a particle inside the nucleus of an atom that is neutral and has no charge.

noncombustible: a substance that will not burn.

noble metal: silver, gold, platinum, and mercury. These are the least reactive metals.

nuclear energy: the heat energy produced as part of the changes that take place in the core, or nucleus, of an element's atoms.

nuclear reactions: reactions that occur in the core, or nucleus of an atom.

nutrients: soluble ions that are essential to life.

octane: one of the substances contained in fuel.

ore: a rock containing enough of a useful substance to make mining it worthwhile.

organic acid: an acid containing carbon and hydrogen.

organic substance: a substance that contains carbon.

osmosis: a process where molecules of a liquid solvent move through a membrane (filter) from a region of low concentration to a region of high concentration of solute.

oxidation: a reaction in which the oxidising agent removes electrons. (Note that oxidising agents do not have to contain oxygen.)

oxide: a compound that includes oxygen and one other element.

oxidise: the process of gaining oxygen. This can be part of a controlled chemical reaction, or it can be the result of exposing a substance to the air, where oxidation (a form of corrosion) will occur slowly, perhaps over months or years.

oxidising agent: a substance that removes electrons from another substance (and therefore is itself reduced).

ozone: a form of oxygen whose molecules contain three atoms of oxygen. Ozone is regarded as a beneficial gas when high in the atmosphere because it blocks ultraviolet rays. It is a harmful gas when breathed in, so low level ozone, which is produced as part of city smog, is regarded as a form of pollution. The ozone layer is the uppermost part of the stratosphere.

pan: the name given to a shallow pond of liquid. Pans are mainly used for separating solutions by evaporation.

patina: a surface coating that develops on metals and protects them from further corrosion.

percolate: to move slowly through the pores of a rock.

period: a row in the Periodic Table.

Periodic Table: a chart organising elements by atomic number and chemical properties into groups and periods.

pesticide: any chemical that is designed to control pests (unwanted organisms) that are harmful to plants or animals.

petroleum: a natural mixture of a range of gases, liquids and solids derived from the decomposed remains of plants and animals.

pH: a measure of the hydrogen ion concentration in a liquid. Neutral is pH 7.0; numbers greater than this are alkaline, smaller numbers are acidic.

phosphor: any material that glows when energized by ultraviolet or electron beams such as in fluorescent tubes and cathode ray tubes. Phosphors, such as phosphorus, emit light after the source of excitation is cut off. This is why they glow in the dark. By contrast, fluorescors, such as fluorite, emit light only while they are being excited by ultraviolet light or an electron beam.

photon: a parcel of light energy.

photosynthesis: the process by which plants use the energy of the Sun to make the compounds they need for life. In photosynthesis, six molecules of carbon dioxide from the air combine with six molecules of water, forming one molecule of glucose (sugar) and releasing six molecules of oxygen back into the atmosphere.

pigment: any solid material used to give a liquid a colour.

placer deposit: a kind of ore body made of a sediment that contains fragments of gold ore eroded from a mother lode and transported by rivers and/or ocean currents.

plastic (material): a carbon-based material consisting of long chains (polymers) of simple molecules. The word plastic is commonly restricted to synthetic polymers.

plastic (property): a material is plastic if it can be made to change shape easily. Plastic materials will remain in the new shape. (Compare with elastic, a property where a material goes back to its original shape.)

plating: adding a thin coat of one material to another to make it resistant to corrosion.

playa: a dried-up lake bed that is covered with salt deposits. From the Spanish word for beach.

poison gas: a form of gas that is used intentionally to produce widespread injury and death. (Many gases are poisonous, which is why many chemical reactions are performed in laboratory fume chambers, but they are a byproduct of a reaction and not intended to cause harm.)

polymer: a compound that is made of long chains by combining molecules (called monomers) as repeating units. ("Poly" means many, "mer" means part).

polymerisation: a chemical reaction in which large numbers of similar molecules arrange themselves into large molecules, usually long chains. This process usually happens when there is a suitable catalyst present. For example, ethene reacts to form polythene in the presence of certain catalysts.

porous: a material containing many small holes or cracks. Quite often the pores are connected, and liquids, such as water or oil, can move through them.

precious metal: silver, gold, platinum, iridium, and palladium. Each is prized for its rarity. This category is the equivalent of precious stones, or gemstones, for minerals.

precipitate: tiny solid particles formed as a result of a chemical reaction between two liquids or gases.

preservative: a substance that prevents the natural organic decay processes from occurring. Many substances can be used safely for this purpose, including sulphites and nitrogen gas.

product: a substance produced by a chemical reaction.

protein: molecules that help to build tissue and bone and therefore make new body cells. Proteins contain amino acids.

proton: a positively charged particle in the nucleus of an atom that balances out the charge of the surrounding electrons

pyrite: "mineral of fire". This name comes from the fact that pyrite (iron sulphide) will give off sparks if struck with a stone.

pyrometallurgy: refining a metal from its ore using heat. A blast furnace or smelter is the main equipment used.

radiation: the exchange of energy with the surroundings through the transmission of waves or particles of energy. Radiation is a form of energy transfer that can happen through space; no intervening medium is required (as would be the case for conduction and convection).

radioactive: a material that emits radiation or particles from the nucleus of its atoms.

radioactive decay: a change in a radioactive element due to loss of mass through radiation. For example uranium decays (changes) to lead.

radioisotope: a shortened version of the phrase radioactive isotope.

radiotracer: a radioactive isotope that is added to a stable, nonradioactive material in order to trace how it moves and its concentration.

reaction: the recombination of two substances using parts of each substance to produce new substances.

reactivity: the tendency of a substance to react with other substances. The term is most widely used in comparing the reactivity of metals. Metals are arranged in a reactivity series.

reagent: a starting material for a reaction.

recycling: the reuse of a material to save the time and energy required to extract new material from the Earth and to conserve non-renewable resources.

redox reaction: a reaction that involves reduction and oxidation.

reducing agent: a substance that gives electrons to another substance. Carbon monoxide is a reducing agent when passed over copper oxide, turning it to copper and producing carbon dioxide gas. Similarly, iron oxide is reduced to iron in a blast furnace. Sulphur dioxide is a reducing agent, used for bleaching bread.

reduction: the removal of oxygen from a substance. See also: oxidation.

refining: separating a mixture into the simpler substances of which it is made. In the case of a rock, it means the extraction of the metal that is mixed up in the rock. In the case of oil it means separating out the fractions of which it is made.

refractive index: the property of a transparent material that controls the angle at which total internal reflection will occur. The greater the refractive index, the more reflective the material will be.

resin: natural or synthetic polymers that can be moulded into solid objects or spun into thread.

rust: the corrosion of iron and steel.

saline: a solution in which most of the dissolved matter is sodium chloride (common salt).

salinisation: the concentration of salts, especially sodium chloride, in the upper layers of a soil due to poor methods of irrigation.

salts: compounds, often involving a metal, that are the reaction products of acids and bases. (Note "salt" is also the common word for sodium chloride, common salt or table salt.)

saponification: the term for a reaction between a fat and a base that produces a soap.

saturated: a state where a liquid can hold no more of a substance. If any more of the substance is added, it will not dissolve.

saturated solution: a solution that holds the maximum possible amount of dissolved material. The amount of material in solution varies with the temperature; cold solutions can hold less dissolved solid material than hot solutions. Gases are more soluble in cold liquids than hot liquids.

sediment: material that settles out at the bottom of a liquid when it is still.

semiconductor: a material of intermediate conductivity. Semiconductor devices often use silicon when they are made as part of diodes, transistors or integrated circuits.

semipermeable membrane: a thin (membrane) of material that acts as a fine sieve, allowing small molecules to pass, but holding large molecules back.

silicate: a compound containing silicon and oxygen (known as silica).

sintering: a process that happens at moderately high temperatures in some compounds. Grains begin to fuse together even through they do not melt. The most widespread example of sintering happens during the firing of clays to make ceramics.

slag: a mixture of substances that are waste products of a furnace. Most slags are composed mainly of silicates.

smelting: roasting a substance in order to extract the metal contained in it.

smog: a mixture of smoke and fog. The term is used to describe city fogs in which there is a large proportion of particulate matter (tiny pieces of carbon from exhausts) and also a high concentration of sulphur and nitrogen gases and probably ozone.

soldering: joining together two pieces of metal using solder, an alloy with a low melting point.

solid: a form of matter where a substance has a definite shape.

soluble: a substance that will readily dissolve in a solvent.

solute: the substance that dissolves in a solution (e.g. sodium chloride in salt water).

solution: a mixture of a liquid and at least one other substance (e.g. salt water). Mixtures can be separated out by physical means, for example by evaporation and cooling.

solvent: the main substance in a solution (e.g. water in salt water).

spontaneous combustion: the effect of a very reactive material beginning to oxidise very quickly and bursting into flame.

stable: able to exist without changing into another substance.

stratosphere: the part of the Earth's atmosphere that lies immediately above the region in which clouds form. It occurs between 12 and 50 km above the Earth's surface.

strong acid: an acid that has completely dissociated (ionised) in water. Mineral acids are strong acids.

sublimation: the change of a substance from solid to gas, or vica versa, without going through a liquid phase.

substance: a type of material, including mixtures.

sulphate: a compound that includes sulphur and oxygen, for example, calcium sulphate or gypsum.

sulphide: a sulphur compound that contains no oxygen.

sulphite: a sulphur compound that contains less oxygen than a sulphate.

surface tension: the force that operates on the surface of a liquid, which makes it act as though it were covered with an invisible elastic film.

suspension: tiny particles suspended in a liquid.

synthetic: does not occur naturally, but has to be manufactured.

tarnish: a coating that develops as a result of the reaction between a metal and substances in the air. The most common form of tarnishing is a very thin transparent oxide coating.

thermonuclear reactions: reactions that occur within atoms due to fusion, releasing an immensely concentrated amount of energy.

thermoplastic: a plastic that will soften, can repeatedly be moulded it into shape on heating and will set into the moulded shape as it cools.

thermoset: a plastic that will set into a moulded shape as it cools, but which cannot be made soft by reheating.

titration: a process of dripping one liquid into another in order to find out the amount needed to cause a neutral solution. An indicator is used to signal change.

toxic: poisonous enough to cause death.

translucent: almost transparent.

transmutation: the change of one element into another.

vapour: the gaseous form of a substance that is normally a liquid. For example, water vapour is the gaseous form of liquid water.

vein: a mineral deposit different from, and usually cutting across, the surrounding rocks. Most mineral and metal-bearing veins are deposits filling fractures. The veins were filled by hot, mineral-rich waters rising upwards from liquid volcanic magma. They are important sources of many metals, such as silver and gold, and also minerals such as gemstones. Veins are usually narrow, and were best suited to hand-mining. They are less exploited in the modern machine age.

viscous: slow moving, syrupy. A liquid that has a low viscosity is said to be mobile.

vitreous: glass-like.

volatile: readily forms a gas.

vulcanisation: forming cross-links between polymer chains to increase the strength of the whole polymer. Rubbers are vulcanised using sulphur when making tyres and other strong materials.

weak acid: an acid that has only partly dissociated (ionised) in water. Most organic acids are weak acids.

weather: a term used by Earth scientists and derived from "weathering", meaning to react with water and gases of the environment.

weathering: the slow natural processes that break down rocks and reduce them to small fragments either by mechanical or chemical means.

welding: fusing two pieces of metal together using heat.

X-rays: a form of very short wave radiation.

Index